THE CHOSEN BODY

Nostalgia Jewishness is a lullaby for old men

gumming soaked white bread.

J. GLADSTEIN, *modernist Yiddish poet*

CONTRAVERSIONS

JEWS AND OTHER DIFFERENCES

DANIEL BOYARIN

CHANA KRONFELD, AND

NAOMI SEIDMAN, EDITORS

The task of "The Science of Judaism"

is to give Judaism a decent burial.

MORITZ STEINSCHNEIDER,

founder of nineteenth-century

philological Jewish Studies

THE CHOSEN BODY

THE POLITICS OF THE BODY IN ISRAELI SOCIETY

MEIRA WEISS

Stanford University Press • *Stanford, California*

Stanford University Press

Stanford, California

© 2002 by the Board of Trustees of the

Leland Stanford Junior University

Printed and bound by CPI Group (UK) Ltd,
Croydon, CR0 4YY

Library of Congress-Cataloging-in-Publication Data

Weiss, Meira

   The chosen body : the politics of the body in Israeli society / Meira Weiss.

     p. cm. — (Contraversions—Jews and Other Differences)

   Includes bibliographical references and index.

   ISBN: 978-0-8047-3272-7

     1. Body, Human—Social aspects—Israel. 2. National characteristics, Israeli.

   3. Jews—Israel—Identity. 4. Halutzim—Social aspects. 5. Sabras—Social aspects.

   6. Body image—Social aspects—Israel. 7. National characteristics, Israeli.

   I. Title. II. Series.

   GT497.I75 W45 2002

   306.4—dc21   2002009776

Original printing 2002

Last figure below indicates year of this printing:

11   10   09   08   07   06   05

Typeset at Stanford University Press in Minion and Copperplate

ACKNOWLEDGMENTS

This book is based on several of my previous studies. It is therefore a pleasant obligation to thank, once again, all those who helped me do research in recent years. I thank my friends and colleagues in the Department of Sociology and Anthropology, the Hebrew University (Jerusalem), as well as my associates in the Israeli Anthropological Association and the Lafer Center for Gender Studies. I was very fortunate to be assisted by a few wonderful graduate students who provided me with thoughtful reflections. I am especially grateful to Elly Teman, Tzipy Ivry, Limor Samimayan, and Michal Lester Levy. I am also indebted to my many students who read and commented on various versions of this manuscript. Special thanks go to the series editors, Daniel Boyarin and Chana Kronfeld, and to the devoted editor Helen Tartar.

The book was written during a sabbatical in the Department of Anthropology at the University of California, Berkeley. I thank my hosts there for providing the conditions and atmosphere that enabled the formation of this book. I am particularly indebted to my hosts and friends Nancy and Michael Scheper-Hughes, as well as to many other professors and students at Berkeley: Stanley Brandes, Paul Rabinow, Lawrence Cohen, Sharon Kaufman, Stefania Pandolfo, Pnina Abir Am, Katarine Young, Eric Stober, Eric Klinenberg, Jody Ranck, and Paco Ferrandiz-Martin Francisco. I am also greatly indebted to the following people for their insights and criticism: Margaret

Lock, Gelya Frank, Emily Martin, Myra Bluebond Langer, Jean Comaroff, Sarah Franklin, Ruth Behar, Lois Wacquant, Paul Willis, Efrat Tseelon, Ruth Tsoffar, Jeanne Katz, Judith Lorber, Debora Gordon, Aliza Kolker, and Matti Bunzel. Special thanks go to the two manuscript reviewers, Ruth Linden and Ronit Lentin. During my visit to Norway I was fortunate to have wonderful meetings with the writer Susan Senstan. Later she became an important reader of my work, and I thank her for that. I wish to thank Ann Louise and Benedicte Ingstad too. In a visit to Turkey I met Aklile Gursoy and her students, whose anthropological insights were very significant in the context of Middle Eastern studies.

I thank my friends and colleagues at the Department of Sociology and Anthropology at the Hebrew University of Jerusalem: Amalya Oliver, Baruch Kimmerling, Boas Shamir, Don Handelman, Don Seemen, Dani Rabinowitz, Daniel Maman, Eyal Ben-Ari, Erik Cohen, Gideon Aran, Harvey Goldberg, Hagar Salmon, Judith Shuval, Michael Shalev, Nachman Ben-Yehuda, Nissim Mizrahi, Orna Sasson-Levy, Sigal Gooldin, Tamar Elor, Tamir Sorek, Vered Vinitzky Seroussi, Victor Azarya, Yoram Bilu, Yehuda Goodman, and Zeeve Rosenhek. I also thank my friends outside the Hebrew University: Dafna Izraeli, Dan Baron, Michael Gluzman, Yehuda Hiss, Yona Tenenbaum, Chen Kugel, and Ilan Goor Arie. My talks with Israeli artists are interwoven through this book. I particularly thank Meir Gal and Yuval Caspi. They stimulated my thinking more than anyone.

Last but definitely not least, I thank my husband Shraga and my children Tami and Shay for their love and support. They were—as always—a source of hope and light. I also thank my beloved family in New York: Gladys and Donald Stone, Shirely and Arnold Bart Elinor, and Bob Rosenblat. I thank my friend from Pennsylvania, Lynn Uram, whose daily phone call to Berkeley made the difference in my writing, and my "foster parents" in Berkeley, Dalia and Shuka Etzion.

This study was supported by generous funds from the Minerva Center for Human Rights and the Scheine Research Center.

I have used several of my previously published studies as sources for secondary analysis conducted for this book's purposes, and they are reproduced with the permission of the publishers: for pages 52–53 of Chapter 2, "War Bodies—Hedonist Bodies: Dialectics of the Collective and the Individual in Israeli Society," *American Ethnologist*, 24, 4 (1997): 1–20; for pages 66–71, Chapter 3, "Bereavement, Commemoration and Collective Identity in Contemporary Israeli Society," *Anthropological Quarterly*, 70, 2 (1997): 91–101; for pages 103–10, Chapter 4, "Engendering the Gulf War: Israeli Nurses and the Discourse of Soldiering," *Journal of Contemporary Ethnography*, 27, 2 (1998): 197–218, and for Chapter 6, "'Writing Culture' Under the Gaze of my Country," *Ethnography*, 2, 1 (Mar. 2001): 77–91.

CONTENTS

INTRODUCTION   *1*

1   THE BODY AS SOCIAL MIRROR   *9*

2   CHOOSING THE BODY: PREGNANCY, BIRTH,
MILITARY, WAR, AND DEATH   *27*

3   SANCTIFYING THE CHOSEN BODY:
BEREAVEMENT AND COMMEMORATION   *65*

4   ENGENDERING THE CHOSEN BODY:
WOMEN AND SOLDIERING   *94*

5   THE CHOSEN BODY AND THE MEDIA   *118*

6   WRITING THE BODY   *135*

NOTES   *145*
REFERENCES   *157*
INDEX   *175*

THE CHOSEN BODY

So, diaper your son with uniforms, give him an anti-pyretic,
anti-innocence military suppository, so that he falls in love
with his penis, with the nation's penis, the army's, so that
he'll be aroused by the smell of gun powder. . . . Hey, cute
victim, what will you be, if you get to be, when you're older?
A man and a parachuter? And later? A grave.

From the poem "Yeru, Yeru, Yerushalayim,"

by Yitzhak La'or (1999)

INTRODUCTION

THE ZIONIST REVOLUTION that aimed to create a new people fit for a new land had a unique bodily aspect. Zionism was not just a national, political, and cultural movement of liberation, but also a bodily revolution. For early Zionist thinkers like Max Nordau and Aharon David Gordon, returning to Israel and working the land would restore the health of Jewish bodies. The Zionist revolution involved a "return" to Zion, to nature, and to the body. Agriculture, land, territory, and military power were seen as an antidote to what was perceived as the passivity and spirituality of Jews and Judaism in the diaspora. The essential feature of the Jewish collectivity, namely, learning, was traditionally disembodied; Zionism regarded itself as a revolutionary attempt to re-embody the Jew, reposition him in history, striving for a worldly redemption. In Nordau's term, coined as early as 1898, Zionism was to be "Judaism with muscles."[1] The "muscle Jew" was to replace the pale-faced and thin-chested "coffeehouse Jew," and to regain the heroism of his forefathers in the land of Zion.

The dictum of "Judaism with muscles" was accepted and elaborated by many passsionate Zionists in the following years. Ze'ev Jabotinsky, the influential leader of right-wing revisionist Zionism, to cite just one of many examples, lamented in his essay "Silk and Steel" (1924) that "the Hebraic man, who has not been taken care of throughout the generations, has been spoiled and become a caricature. Bent backbone, pointed chin, low lip—

These are his characteristics. You [Zionist mothers of Israel] will rebuild the generation." Doctor Binyamini, house physician of the famous Herzliya Gymnasium, the first Hebrew school in Tel Aviv, wrote in his 1928 diary that "Zionism was accepted only by compatible men and women who were whole-bodied and physically fit. . . . Our people are currently experiencing a natural process of selection." Doctor Rupin, head of the Israeli Office of the World Zionist Organization, wrote in his book *The Sociology of Jews* (1934) that "while in Europe many are calling for a eugenic policy, the Jews . . . have never engaged in a 'self-cleansing' of their race, but rather allowed every child, be it the most sickly, to grow and marry and have children like him. Even the mentally retarded, blind and deaf were allowed to marry. In order to keep the purity of our race, such Jews must abstain from child-bearing."

From the 1920's on, the healthy and happy baby, "little Samson" as he was called in the posters of the day, figured regularly in mainstream Hebrew literature on parenting and child care, where he served as an antithesis to the degeneration of the diaspora. Women were defined, following 1948, as responsible for bearing as many children as possible "for the glory of the State of Israel." For many women, the revolutionary turn of Zionism therefore culminated in a return to a traditional gender role, namely, mothering.

Today, Israeli society is still obsessed with fertility. The prestate zeal for community eugenics and the post-independence craving for quantity and quality have found their contemporary consummation in genetic screening, testing, and counseling. Israeli women hold the world record for fetal diagnostics. About 60 percent of pregnant women in Israel undergo some kind of diagnostic test before delivery. Israel has the world's largest number, on average, of tests per pregnancy. It is the first in the world in the rate of amniocentesis (undertaken by about 20 percent of pregnant women).[2] In addition, Israel has more fertility clinics per capita than any other country in the world, not because of high rates of infertility but because of the centrality of reproduction. Finally, Israeli parents are prone to prefer abortion even in cases of relatively mild impairments, such as cleft lip, and in cases of relatively low risk, such as 5 percent risk of mental retardation. This shows that the Israeli obsession with fertility involves not just quantity but also quality.

All this reproductive activity revolves around the birth of a healthy baby. The many tests conducted before and during pregnancy are meant to guarantee a perfect child. Contemporary Israelis, according to geneticists, are the most enthusiastic about new prenatal diagnostics, and the first to spend money on them.[3] Every new test is immediately put into use, with the result that the number of administered prenatal tests increases threefold each year. Israeli geneticists no doubt have a vested interest in this popular medical industry. Their attitudes toward abortion are, on average, more positive and accepting than those of their international colleagues. For example, fully 68 percent of them agree that giving birth to a child with a serious impairment is socially wrong. In contrast, geneticists around the world usually regard the decision to abort a deformed fetus as primarily personal. An astounding 14 percent of the Israeli geneticists agree that "the role of genetics is to purify the human genetic pool" (Wertz 1998). Such a view is ethically problematic, indicating as it does a geneticist's directive involvement in the decision-making process of individuals. It flouts the international code of ethics of genetic counseling, which stresses non-directiveness. Representatives of international adoption agencies agree that Israelis are the most critical in the world when it comes to considering a candidate for adoption and would refuse to adopt a child because of minor external impairments.[4] Eugenics, then and now, is a social discourse that has gained prominence in Israel through the ideology of the "chosen body."

What is at issue here is the ideology of the chosen body in Israel, an idea that has been long and consistently cherished by Israeli society and, as some argue, also by Israeli sociology, which has always taken an active part in the Zionist project of nation-building. The same themes, in contrast, have been sharply criticized by Jewish American observers like David Biale, Daniel Boyarin, and Howard Eilberg-Schwartz. Biale (1992a), for example, opens his book *Eros and the Jews* with a discussion of Philip Roth's notorious 1969 novel, *Portnoy's Complaint*, "the outrageous and hilarious confession of a sex-obsessed American Jew." Biale reminds us that, for Roth, the Jews as "the quintessential People of the Book" live in symbolic exile from their own body. At the end of the novel, Portnoy therefore chooses to escape to the healthy sexuality of the Jewish state and seeks an affair with a sturdy female Israeli soldier. Roth combines here two powerful elements in the myth of modern Israel: Eros and the military.

Portnoy's object of pursuit, however, turns out to be depressingly puritanical. This should remind us that "erotic Israel" is a sexual stereotype in American Jewish culture. For commentators like Biale, Boyarin, and Eilberg-Schwartz, the Zionist Eros is not an aberration of history but the crystallization of sexual themes long inherent in Judaism. Judaism is also about "the people of the body" (Eilberg-Schwartz 1992), not just "the people of the book," and about "carnal Israel" seen through its Talmudic texts (Boyarin 1993). I intend to follow these themes and discuss the continuities and discontinuities between the "book" and the "body," the Jewish Eros and the Israeli Eros, as they are embedded in the attempt to construct a new person in the Jewish state.[5] Georges Friedmann, the eminent French sociologist, concluded in his book *The End of the Jewish People* that "a new people is being created every day in Israel; a young people that is neither an appendage nor the center of the now-legendary 'Jewish people'" (1967: 238). This book examines the discursive creation of that new people through the prism of the chosen body.

The claim that bodies are the concrete articulations of abstract social paradigms has been the impetus behind a recent upsurge of scholarly literature in medical anthropology, the sociology of the body, gender studies, cultural studies, and many other related fields. This book offers to examine how the social paradigms of contemporary Israel are articulated through the body. It examines Israeli society as a case study in the sociology of the body. To gain a panoramic view of how the body is regulated, cared for, chosen, and ultimately made perfect and immortalized, the book brings together different analyses of several ethnographic fields. They are the fruit of several studies that I carried out during more than 20 years of academic research; all of them focus on the body, and all of them are joined here together for the first time within one overarching theoretical framework.[6]

Several researchers have already probed into the connection between body-building and nation-building. Anita Shapira, for example, wrote that Zionism sought to "replace the *Yeshiva* (religious highschool) student with a healthy youngster, daring and ready for battle." Michael Gluzman (1997) emphasizes the gendered aspect of that conversion.[7] In his words, the Zionist body, Jewish, Ashkenazi, and male, is "Maccabi reincarnated." This trope appears, for example, in canonical Zionist texts such as Herzl's *Alteneuland* (Old-New Land, meaning Zion). Gluzman and others have stressed the gendered aspects of the "pioneer" and the "sabra," yet ignored another, no less important physical characteristic: the idealization of health, power, and perfection. This masculine, Jewish, Ashkenazi, perfect, and wholesome trope is what I call, for short, the chosen body. It is an ideal type by which concrete Israeli bodies are screened and molded from their birth to their death.[8]

My theoretical point of departure is that Israeli society (like the prestate community before it) has always molded and regulated bodies as part of the ongoing construction of its collective identity. Israel's continuing involvement in an armed conflict with its Arab neighbors has become linked to a society deeply concerned with territorial borders, as well as body boundaries. Since the early days of nation-building (1900's–40's), the Israeli/Zionist body has been regulated to form a "new person."

To be sure, collectivist embodiments of chosen bodies are not unique to Israeli society. But Israeli culture has left its unique mark on two specific embodiments, notably, the "pioneer" (halutz) and the "sabra." These two mythical figures—the prestate pioneer and the nation-building and post–1948 sabra—share a family resemblance. Both embody standardization, self-mortification, and a career in the service of the nation. Both were (and are) used by the collectivity for purposes of inclusion and exclusion, since no Arab or other non-Jew is capable of being a pioneer or a sabra. Like the social institution of pioneering, Israeli soldiering also defined the boundaries of the new Jewish collectivity, since Israeli Arabs were (and are) excluded from military service.

I discuss the Israeli pioneer, and his American Fordist counterpart, in the following chapter, which deals with the historical background of the politics of the body and the body politic in Israel. The body in its gendered, utopian, and collectivist construction, that is to say, is used there as a "backstage window" into the crystallizing Israeli collectivity. Collective embodiments, such as the pioneer and the sabra, provide a powerful analytical gateway into the dominant discourse. As a discourse, collectivism often hides its socializing program, seductively obscuring it as taken for granted, beyond refutation and criticism. From Zionist thinkers to Israeli citizens, collectivism has long been perceived not as a threat to the autonomy of the individual but rather as an emancipating force. The basic Jewish-Israeli experience during the twentieth century encompassed the transition from a state of dependency and dispersion in the diaspora to a state of sovereignty backed by military and national power, which is therefore perceived as emancipatory. In a similar way, the Zionist pioneer and the Israeli sabra are often perceived as singular and individualistic, although in practice both are the agents of Zionist collectivism. In hindsight, we can say that a collectivist trope such as the Zionist pioneer was cleverly tailored to carry an individual message so as to make it more attractive. In Marxist terms, pioneering was also a form of "false consciousness" covering normative and ideological control over the individual. The message was that self-fulfillment depends on fulfilling the national goals.

The conquering of land and labor during the first *aliyot* (waves of immigration to Palestine), the struggle for survival and independence in 1948, and the succeeding military conflict-cum-routine were all used to shape, justify, and sustain the construction of the Israeli body through a public-private linkage that is unfamiliar in Western democracies. Collectivism became the "civil religion" of Israel, the larger frame of reference through which other issues and problems—such as militarization, the melting pot of immigration, the relations with the diaspora Jews and the Palestinians—are all defined and accounted for. This emphasis no doubt also had its functional reasons, which, however, do not make it any less subjugating. Its embodiments—such as the pioneer and the sabra—are the visible traces of the collectivity's invisible hand.

A word about the structure of the book. Chapter One discusses the politics of the Israeli body by introducing early Israeli imagery of the body while connecting it to its formative social processes. The logic of presentation follows closely Emily Martin's (1990, 1994) articulations of body imagery vis-à-vis social processes. This book, however, does not directly support or refute her thesis of a turn from collectivism to individualism. My argument is that even if such a turn has taken place—and this should be carefully studied in the local context—the chosen body still exists alongside that change and provides the connective tissue between different ethnic and religious groups in Israel. Moreover, it seems to me that collectivism is still the larger frame of reference in which grow variegated, and perhaps short-lived, cases of individualism.[9] The predominance of the collectivity is connected to the military threat Israel faces on a daily basis (whether through terrorism or war). The stronghold of collectivism and the military conflict-cum-routine are tied together in a vicious circle.

I discuss the Israeli militaristic script, its body imagery, and social repercussions in Chapter Two. This chapter brings together two seemingly disparate areas—"conditional parenting" and soldiers' screening—within the perspective of body regulation. In the context of reproduction and child care, I examine Israeli parents' attitudes toward the body of their fetus and their newborn. As part of this examination, I employ a "second reading" of my book *Conditional Love* (1994), which dealt with the appearance-impaired child in Israel. From there I proceed to the culture of the chosen body in the screening of combat soldiers. An independent section of the chapter adds a comparison between the regulation of impaired bodies of children and the regulation of diseased bodies of grown-ups. The aim of this section is to portray the difference between the collectively regulated bodies of children and soldiers, and the lack of such collective regulation in the case of chronic disease. The chapter ends with a discussion of the chosen body and the Institute of Forensic Medicine.

Chapter Three locates the chosen body in what can be seen as the unfortunate yet natural outcome of militarization, namely, bereavement and commemoration, practices that have a special significance in Israeli society. In a broader sense, I am concerned to show how national ideologies of commemoration construct a narrative of embodiment through which (the corpses of) fallen soldiers become symbols and guardians of some collective body. The chapter examines the "body politic" of commemorative practices and then moves on to describe the semiofficial and individual reactions of bereaved parents and relatives. The concluding section offers another reading of the observations presented in the preceding chapters. My aim is to link the chapters together by asking how the body connects the new Israeli collective identity to the traditional Jewish collective identity and how it separates them. Chapter Four consists of three sections, all pertaining to the issue of gender and the chosen body. The first section analyzes images of the chosen body in the war poetry of Israeli men and women; the second complements the analysis of poetic narrative with interviews with fathers and mothers of combat soldiers; and the third discusses the participation of Israeli women in the masculine ethos of the body from a different, occupational point of view. I focus on the field research I conducted during the Gulf War in order to describe the construction of "new soldiers": the nurses who, upon hearing the SCUD alarm sirens, left their families in their homes and headed toward the hospital. This role reversal (typically the men are the ones who leave their families behind and go to the front) illustrates a unique attempt by women to appropriate the men's ethos of soldiering.

Chapter Five, which concludes the study as such, concerns itself with the media and the discourse that emerges following threats on the public body, such as terrorist bombings, military accidents, and political assassinations. These events, of which Israel has unfortunately recently had more than its share, are all triggers of collective feelings hovering between solidarity and conflict. The construction of these media events is immersed in bodyTalk—the rhetorical use of the body, especially the chosen body, as a mediator of public and private feelings.

In Chapter Six, the Conclusion, I consider the book's thesis from a more personal point of view. But, as in the paragraph below, glimpses of personal reflection are also scattered in relevant places throughout the book. These italicized comments interrupt, reflect on, and argue with the text.[10] Together, they constitute an experiment in "biographical positioning" (Richardson 1997: 107). I have tried to write about myself without essentializing or valorizing, as Laurel Richardson suggests. My biography is important in the context of my research. I will use it to illustrate the analogies between my life and the life of my respondents.

I was born in Israel, to European Jews who came to Israel when they were children. At the encouragement of his socialist youth movement, Dad immediately joined a kibbutz. During the Second World War, he served in the British army. Mother belonged to the right-wing resistance, Etzel, which was very much against the theory and practice of Dad's left-wing youth movement. I was a child from an upper-middle-class, white background, who became an officer in the Israeli military, and later a professor at the Hebrew University, and a mother of children who also served in the military. My life course represents all the "right choices," yet it was also a course chosen for me. Through the biographical positioning that ensues, I try to describe how the Israeli politics of the body has influenced me and my family.

1 THE BODY AS SOCIAL MIRROR

THE SOCIAL REGULATION of the body exists in all human societies. In simple societies, it usually takes the form of explicit rules that result in punishment when broken. In modern societies, the regulation is more subtle, camouflaged in ways that make people believe they regulate their bodies out of their own free will. But in fact modern societies contain many corporeal regulations reflecting class structure and cultural fashions.

Collectivism is far more explicit in its regimes of bodily control. In earlier times, the state (or the monarchy) typically reaffirmed its authority by subjecting its enemies to public torture and death. Modern consumer culture has moved such barbaric structures of authority to the background. The "hedonistic" body, however, is still socially and self-regulated to comply with collective fads, fashions, and lifestyles. Contemporary consumer culture operates through the dramaturgy of the body, and glorifies the virtues of beauty, fitness, and above all youth. Such values are internalized and become "voluntary lifestyles," although they are socially prescribed and usually not to our individual comfort.[1]

Because early social scientists traditionally treated the body as a universal biological entity that "[fell] naturally into the domain of the basic sciences and [was] therefore beyond the purview of social and cultural anthropology" (Lock 1993: 134), the body was not a central subject of research in the social sciences until relatively late in the twentieth century. Since the late

1970's, as Lock (1993) argues in her review of the body in anthropology, bodily practices and knowledge have come under sharp sociological scrutiny.[2] By 1986, as J. M. Berthelot (1986) has remarked in regard to socio-anthropological literature, "the body would now appear to be everywhere." Of course, the body has always been everywhere; it is exactly this pervasiveness that makes it socially important. But the social sciences were very slow to start to look at the different places where the body is found.

Broadly defined, the sociological interest in the body falls into three distinct categories: the physical/individual body, the social body, and the body politic.[3] The first stage of looking at the body stressed its physical aspects. The traditional view of the body as a natural and universal *tabula rasa* arguably emerged from the Western opposition between nature and culture (Rorty 1986). The body was seen as part of nature and hence outside of culture. At the end of the nineteenth century, for example, Emile Durkheim (cited in Lock 1993: 135) wrote that "man is double," distinguishing between the universal, physical body and the "higher" morally imbued "socialized" self that was to be the subject matter of sociology. In a similar way, the body was defined as "the first and most natural tool of man" by the *Annales* group (cited ibid.). The body, therefore, had to be made "social" before embodiment could come into theoretical being.

The anthropologist Arnold Van Gennep played a paramount role in bringing the body back into the social world. In his comparative research on the rites of passage (1908), Van Gennep gave the body a prominent role as the tool and indicator of transition. The body, for example, was mortified and sometimes mutilated in order to signal a successful change of status. Van Gennep and the anthropologists of that time, however, still saw the rite of passage, in the Frazerian "out-of-context" anthropological tradition, as a universal, evolutionary scheme (see Strathern 1987: 254; Gellner 1985: 645; and Lienhardt 1966: 27). The body was assigned a passive and similar role in all cultures. This view was later developed in structural-functional anthropology, especially by Mary Douglas (1966, 1973), whose analysis of the body among the ancient Hebrews will be considered below. It was only in the 1960's and onward that the body and its rites of passage became "embodied" in local cultures, usually counter-cultures. The work of Michel Foucault (1971, 1973, 1977, 1980) further placed the body in a political context, defining the "body politic" as "a set of material elements and techniques that serve as weapons, relays, communication routes, and supports for the power and knowledge relations that invest human bodies and subjugate them by turning them into 'docile bodies'" (Foucault 1977, cited in Rabinow 1984: 176).

Following Foucault, the body now came to be seen within a political context, as an object whose meaning is contingent on processes of knowledge and power. This critical view was also developed by sociologists working in the Marxist and feminist traditions, who stressed the role of the body as a window to class and gender hierarchies. "Writing the body" (Game 1991) has now become a sort of feminist ideological platform signifying an alternative mode of "embodied thought," a deconstruction of the male logocentric master narrative of rationality (Braidoti 1994).[4] "Writing the body (politic)" hence becomes a subversive strategy, capable of "undoing" the social.[5] This motivation is also behind the present study, which sets out to undo the political construction of the body in Israeli society. I also try to "write the body," namely, to use it as an Archimedian point from which malestream ideology (to use the apt feminist jargon) can be deconstructed and things that have become second nature for Israelis can be exposed.[6]

Today, the three bodies—the physical, the social, and the political—once constituted as part of different sociological agendas, are interconnected. The school of cultural studies that has developed since the 1980's around the British journal *Theory, Culture, and Society* sees the body as a synthetic and performative concept that necessarily involves all the aforementioned research traditions (Featherstone et al. 1995; Frank 1990). Indeed, the wide-ranging field of "embodiment" has recently come to be used to bridge the gap between naturalism (stressing the physical/individual body) and constructionism (emphasizing the social/political body). As a result, however, "embodiment" is also becoming a discourse—a regime of signification—and less a strictly defined analytical concept. It is possible that embodiment has already become too vague a concept; future studies will undoubtedly explore this open question.

Some scholars suggest that sociologists today must reconcile the polarities of the "dual body," the body as either a natural/phenomenological entity and/or the product of discourse (see, for example, Hallam et al. 1999: 6). Following Foucault, Chris Shilling (1993) has described a possible way of reconciling the physical/phenomenological and the social/political by framing the body as "an entity which is in the process of becoming." In the postmodern era, the body is a "project"; we find ourselves continuously "working on" the body, and this work practically never ends. The body in the post-modern era is always missing something. In his or her attempt to fill this lack, the individual can perhaps come closer to an ideal of the chosen body. This definition echoes with religious associations; but the only religion in this context is "body religion." Arthur Frank (1990) spoke about this when referring, following Foucault, to the "disciplined body." His definition fits well within this study. If bodies are always in the process of becoming, always in the process of being finished, then individuals are faced with multiple choices. These choices are shaped by cultural scripts, one of the most notable being militarism. By providing soldiers with clear and disciplinary bodily requirements, the military allows them to "complete" their corporeal deficiency and find their prescribed military identity. As Foucault (1980: 59) wrote, "If it has been possible to constitute a knowledge of the body, this has been by way of an ensemble of military and educational disciplines." The military, in a manner parallel to other social institutions, uses the body as a template on which to imprint signals that identify the person with a certain group and exclude him from other groups. In this process, which Bourdieu (1977) termed "somatization," the body becomes at once an object of socialization and the subject of labor and social activity.

My own analysis of the chosen body in Israeli society focuses on these choices, and combines the physical, the social, and the political. I will be concerned with representations of the body—or embodiments—as discursive formations, mediated through language as well as ingrained practices. These representations are constructed both in interviews with people who described their felt experience to me and in public texts, documents, and larger cultural scripts.

The remaining part of this theoretical exposition deals with one outstanding contribution to the recent sociology of the body in order to juxtapose it with my own analysis of Israeli society. The essay I have in mind is Emily Martin's "The End of the Body" (1990). I also draw heavily on her later book, *Flexible Bodies: The Role of Immunity in American Culture from the Days of Polio to the Age of AIDS* (1994).

In her lucid description of the transformations of the body in the natural sciences, Martin defines two periods of changing "paradigms" in American society. First, there is what she calls the paradigm of the Fordist body, which is embodied in the images of reproductive biology: bodies organized around principles of centralized control and factory-based production

(1990; see also Martin 1987, 1991). In this worldview, "men continuously produce wonderfully astonishing quantities of highly valued sperm, women produce eggs and babies, and, when they are not doing this, either produce scrap (menstruation) or undergo a complete breakdown of central control (menopause)" (Martin 1990: 121).

This early-twentieth-century paradigm was gradually transformed into a "late capitalism" paradigm that began in the 1970's. In this new paradigm of flexible specialization, time and space are compressed, capital flows across all borders, and all points are interconnected by instantaneous communications.[7] According to Marshal McLuhan's vision, these far-reaching technological changes should have transformed the world into a "global village" where consumerist culture replaces previous national borders. However, as illustrated by Bosnia, Chechnya, Jammu and Kashmir, or Burundi and Rwanda, some borders and national conflicts do not disappear so easily. For Martin, this new paradigm, associated with the political economy of late capitalism, is "embodied" in the discourse of immunology, where the body is depicted as a whole, interconnected system complete unto itself, "an engineered communications system, ordered by a fluid and dispersed command-control-intelligence network" (Haraway 1989: 12). Unlike the assembly-line standardization of the Fordist body, late-capitalist bodies are characterized by "tailor-made specificities," symbolically depicted, in the immunological discourse of the body, in the variable, learned, and context-dependent antibody reaction of the immune system.

Martin's thesis, like any grand historical narrative, has its inaccuracies. These include an inconsistent time perspective (reproductive biology evolved later than the early twentieth century and persisted well after) and a lack of detailed reference to the discipline's own historical development (immunology itself has changed from a "clonal screening hypothesis" to a "cognitive paradigm"; see I. Cohen 1992). Indeed, Martin herself has elaborated on this last point in her *Flexible Bodies* book, which traces the internal changes of American immunology, as well as its popular representations, within the above-mentioned narrative of social change (see especially pp. 40–41).

My general interest in Martin's thesis stems from its overall theme regarding the reflection of social paradigms in body imagery—in other words, the "body politic" behind the private, normal body, and the dialectics between the two.[8] More specifically, the shift in the social order that

Martin argues for is general enough to apply to, or serve as, an overarching framework for discussing the Israeli context. What Martin and other socio-anthropologists show us is that all cultures and societies can be examined through the body. This is indeed a basic premise of all body research in the discipline. As we will later see, Israeli society is unique in using the body as a key metaphor that I analyze through the idiom of the chosen body. The shift that Martin is illustrating through her data on body imagery has been described by political economists as a shift from "standardized mass production" to "flexible specialization." Sociologists have suggested a very similar shift from collectivism to individualism in Israel. My argument is that these two shifts are interconnected.

Fordism was a political economy suitable for a collectivist society. Late capitalism, in contrast, was developed as part of the individualist and open consumer society. Late capitalism therefore encouraged the illusion of the individual as a free agent, where Fordism regarded the individual as part of the social machine. In late capitalism, the individual is supposed to create him- or herself through the body.[9] The parallel social change underpinning different societies such as the United States and Israel is probably due at least in part to similar conditions of economic change. In any event, even though the process may be similar on a general level, in practice each society experiences it differently. The Israeli images of the chosen body are local and unique, and should be read against the local Israeli backdrop.

Gendered Reading of the Chosen Body

Despite the ideological emphasis on egalitarianism through the espoused ethos of Zionism, in practice gender did matter. Those gender differences have become a paramount issue for the recent generation of feminist scholars working in Israel.[10] Most, in their feminist critiques of the status of women in Israeli society, dwell on the role of Jewish law and the state's military policy as tools of oppression; they recognize and highlight the collusive relationship between Judaism and Zionism, and its interdependent relationship with militarism. Despite universal conscription, the military remains exclusively male, and its sex segregation affects women's advancement in Israeli politics, where a career in the military is often a means of advancement.[11]

My argument regarding the chosen body in Israeli society should be extended to include this gendered reading. In other words, despite the public professions of Zionist leaders that the chosen Israeli body (i.e., the "pioneer," or *halutz*, and the "sabra") was genderless, in fact it was always masculine. The masculinization of the New Hebrew became one of the leitmotifs of Zionism and later served as the breeding ground for the Israeli culture of the chosen body. Herzl's *Altneuland* (a portrait of Zion as an Old-New Land) can be read not only as a political utopia, but also as an attempt to cure the psycho-physical illness of the feminine diaspora Jew (Gluzman 1995: 4; J. Boyarin 1997). In that sense, *Altneuland* was the forerunner of *Portnoy's Complaint*. It is no coincidence that *Altneuland*'s protagonist, the melancholic and effeminate German Jew, Dr. Friedrich Lowenberg, had to be taken out of Vienna, the capital of European decadence, and brought to deserted Palestine before he could discover his manhood. It is only in Palestine that Lowenberg becomes engaged to a Jewish girl, Miriam, who frees him of his lifelong homoerotic ties.

Palestine is therefore framed as the necessary site of healthy heterosexual transformation, the only cure for the pathological (melancholic, effeminate) diaspora Jew. The successful heterosexual transformation of the Jewish people can take place only in the land of Palestine. A people without a land, according to Zionism, is a weak, effeminate, impotent people. In his diary, Herzl further describes his thoughts about the suitable education for the new Hebraic man in Zion:

> For twenty years, until they become aware of it, I must educate the boys as military men. . . . But the others too, I will educate to become free and courageous men, so that in times of ordeal they will join as [military] reserves. Education through songs of the homeland, of the Hashmonites, religion, plays, heroes in the theater, honor, etc. (Cited in Gluzman 1997: 149)

Education for manhood is here framed within the European-style military education. The mainstream ideological writings of Zionism were therefore directed to the Jewish man, portraying his "before" (diaspora femininity and weakness) and "after" (Palestine-Israeli masculinity and might). Ya'acov Cahan (1911: 10) described the "new Hebraic man" (*ha'ivri he'chadash*) as "wonderful in his proud march on the land of his ancestors, with the skies of the Lord fresh and pure above his head. Proud and mighty he will march forward, like the ancient Hebraic man."

The Hebrew language differentiates between the sexes, and my translation of this excerpt is faithful to the sexism in the writing. Like the Hebraic man, the "ancestors" is a male form in the original. The passage therefore specifically addresses Jewish masculinity. Women are blatantly missing, since Zionism was primarily conceived by men and for men, as a means for the reconstruction of manhood.[12] The discrimination of gender in practice while keeping an ideological facade of gender equality continued in the passage from pioneers to sabras.

Political Reading of the Chosen Body

I have already discussed the gender aspect of my argument regarding the chosen body and will here concentrate on its broader political reading. It is a reading that takes into account the various historical influences described before—collectivism, pioneering, gender segregation, and militarization—and recasts them in the more recent reality of the military occupation. There is evidently a linkage between militarization, occupation, and chosen body strategies. Throughout this book, I discuss the topic of embodiment; on the political level, there is no more blatant example of embodiment than that of military occupation. The occupied territories, particularly the Gaza strip and "Judea and Samaria," which were occupied in the 1967 war, have been "embodied" within Israeli territory. The body of the nation has engulfed these territories, practically appropriating them while excluding their Arab inhabitants. A whole symbolic array of sacred biblical images was invoked by the "followers of the whole land of Israel" (*Eretz-Yisrael Ha-shlema*) in order to justify and seal this territorial inclusion. With the embodiment of the occupied territories, Israeli society has become an occupying force; Israelis have assumed the role of the oppressor, the overlord—a role that has had an accumulating impact on the country.

This study brings together various examples of the Israeli embodiment of chosen, Jewish, male bodies. Implicit in all these examples is also the disembodiment of "the Other within us," namely, Arabs and Palestinians, but also Sephardim, Ethiopians, Russians, Yemenis, the disabled, the bereaved, lesbians and gays, and so forth. Throughout this book, my use of "Israeli society" refers to Israeli *Jewish* society. My argument could therefore be extended through a political reading of the chosen body as an apt trope for a military-masculine, settler-colonial society. Several Israeli sociologists have recently taken such a tack. According to a historical survey by Uri Ram (1993b), although the colonial studies approach made its first appearance as a specific scholarly project in Israeli academia in the wake of the 1967 war, articles supporting this approach did not begin to appear until much later (at least in English). Gershon Shafir (1996), one of its leading supporters, was among the first to tackle the subject. He argues that the most significant determinant of past and present Arab-Israeli relations was the early Zionist land and labor structures, encompassing a homogeneous form of colonization (see also Grinberg 1994).

On this view, nation-building practices such as the conquest of the land (*kibush ha'karka*) and the conquest of labor *(kibush ha'avoda)* became interwoven with pioneering and the making of a chosen body fit for the chosen land. That is why Israel has such a difficult time disconnecting from its occupied Palestinian land as well as its Palestinian labor force. Although Israeli decolonization would facilitate, at least according to Shafir, the security and autonomy sought by both Israelis and the Palestinians, the separation of the two groups is highly complicated.

Nahla Abdo and Nira Yuval-Davis (1995) have also looked at what they term the "Zionist settler project" (ZSP) in Israel and Palestine. The hegemonic Zionist ideology portrays Palestine as a history-less region of sparse population, seemingly waiting for the return of the Jewish people. In reality, these authors argue, the ZSP was a colonial venture into a region populated by indigenous Palestinians who resented the settlers' intrusion. It was the Zionist settler project (under this or a more euphemistic title) that propelled the ideologies and practices connected with the chosen body, such as militarism, masculinity, collectivism, and subjugation.

Israel's colonization policy was constituted within a larger discourse of incorporation. In what Shafir and his co-author Yoav Peled (1998) call an "incorporation regime," Israel forged its relationship with its main ethnic and religious groups—*ashkenazim, mizrachim*, orthodox Jews, Israeli Arabs, and noncitizen Palestinians—through a hierarchical combination of three citizenship discourses: (1) a collectivist republican discourse, based on the civic virtue of pioneering colonization; (2) an ethno-nationalist discourse, based on Jewish descent; and (3) an individualist liberal discourse, based on civic criteria of membership. The authors suggest that Israel's historical trajectory has consisted of a gradual transformation from a colonial to a civil society and, concomitantly, the gradual replacement of a republican citizenship discourse with a liberal discourse. In an earlier article (1996), Shafir and Peled similarly argue that Israel is moving from an exclusionary frontier society formed in 1948 to a democratic civil society. For years, Labor Zionism's masking of exclusionary colonial citizenship practices within a universalistic discourse prevailed, as did Israel's lack of will to solve the Palestinian crisis. However, the Ashkenazi elite gradually became more globally oriented, and Israeli society opened up to international economic, political, and cultural influences.

From Pioneers to Sabras: Embodiments of the
Chosen Body in Israeli Society

The "chosen body" represents a "deep structure" underpinning Israeli society.[13] Its pervasive expressions can be found in the Zionist ethos of pioneering, as well as in the Jewish religious code (halacha). It is present in the nation-building project after 1948, with the absorption of mass immigration, and in the ongoing construction of collective identity vis-à-vis the excluded others. It continues to play an influential role in cultural scripts such as militarism and nationalism. Furthermore, the chosen body continues to underpin contemporary social change, such as the transformation from collectivism to individualism. Although many sociologists have described such a process in Israeli society, I argue that the grip of the chosen body has not been diminished because of the (putative) turn to individualism. There are three main arguments supporting this claim.

First of all, militarism and nationalism are still at the formative core of Israeli identity. A glance at Israel's power elite would show the interlocking of the military and the political spheres, with generals still being forwarded to lead political parties and head governments. Both militarism and nationalism are tightly connected to the chosen body. Second, if a shift from collectivism to individualism has occurred in Israeli society, it has been gradual and complex. The pioneer and the sabra, whose blueprint was formed in the 1930's–50's, are still an important part of Israel's social reality. Third, although individualism may be superficially regarded through the image of self-identity, it still encompasses a strong "language of commitment" (Bellah et al. 1985). The collectivist "language of commitment" is particularly evident in events that threaten the collective body, such as terrorist attacks, wars, Intifada, and political assassinations (notably Rabin's). Regrettably, there is no shortage of such events in Israeli life. I will discuss some of them in more detail in later chapters. Even when these events are resolved, and life seems to regain its normal course, we still need to look critically at the "individualism" that arguably characterizes Israeli society. As I argued, the fashions and lifestyles of today's individualistic society are in fact collective patterns of consumption. In the Israeli context, individualism still has a negative connotation, while collectivism is generally perceived in a positive manner. Although many Israelis may feel that collectivism was part of their youth (and of youth generally), they perceive individualism—accompanied by "cynicism and fatigue"—not as a real, valid alternative, but as a kind of midlife crisis. "Israel," wrote Amos Elon more than 25 years ago (1971: 322), "resembles a man racing ahead with his eyes turned back."

The tropes of the pioneer and the sabra were cleverly packaged to appear individualistic, while in fact representing a collective practice. There is a unique cultural script at work here. In America, the pioneer ethos stressed individuality, daring, and go-gettism; in modern Hebrew, "halutz" connotes above all service to an abstract idea, to a political movement, and to the community (see Elon 1971: 112). Israel's collectivist ideology in its formative years can be related to the communal utopia of socialist Zionism (Even-Zohar 1981) as well as to the Ashkenazi tradition of communal life: state as shtetle (Liebman & Don-Yehia 1983). In the years of the Yishuv (the Jewish community in Palestine under the British Mandate, 1917–48) and afterwards, the collectivist ideology was the interconnecting link between different origins, occupations, and political affiliations. The view of the individual as the carrier of collective ideals and as subordinate to them was a common thread characterizing the otherwise antagonistic camps of the left-wing pioneers and the right-wing nationalists. Both ideological approaches demanded of individuals that they be prepared to sacrifice their personal interests and put themselves at the service of the collectivity. The primary manifestations of such service in the case of the Labor movement were the kibbutzim (Lissak & Horowitz 1989: 110–11).

Collective action is geared to producing large quantities of standardized products put together from standardized components. The components of the collective are human beings, and these also must be standardized in terms of values, expectations, commitment, and prestige. The internalizing of such a process of standardization can be so far-reaching as to result in the subject's genuine need to fulfill the "requests" of the collective. We can see a prime example of this in Foucault's instance of the Church subjugating a believer by turning him into a "confessional animal." A similar case in point is the social construction of the Israeli pioneer, who is—on the face of it—an individual striving for personal fulfillment. Yet what he (or she) really fulfills is a collective practice. In practice, the pioneer sacrifices his or her personal aspirations on the altar of collectivism, which means the denial of individualism. Indeed, it was this collective conception of the pioneer that underlay the segregating of Jews and Arabs in Palestine, for there were no Arab pioneers, only Jewish ones.

The Israeli pioneer should be seen as the embodiment of the Zionist vision of the New Jew (who is always a male, in Hebrew). Martin Buber, for example, argued in 1936 that the Zionist movement had "proved its authenticity in the image of the *halutz*" (Buber 1961: 255). The construction of the halutz as a new type of person should therefore be located in the history of Zionist thought. The "pioneer" and his predecessor, the sabra, are both "body persons," collectivistic embodiments with carnal rather than intellectual features. Like other late-nineteenth-century nationalists, the Zionists were preoccupied with the physical degeneration of the people. The solution for the diaspora required a homeland; the cure for the disturbed soul demanded exercise and physical labor. The individual body became a microcosm of the national body politic (see Biale 1992a: 178–79). The dichotomy between the corporal and the spiritual became widely recognized as the essential Jewish condition and predicament.

One of the more prolific turn-of-the-century Hebrew writers, the social philosopher Ahad Ha'am,[14] identified these two elements in his essay "Flesh and Spirit." He argued that following the destruction of the Second Commonwealth, these two elements became differentiated. Judaism, having to exist without the material infrastructure of a body politic, became predominantly spiritual. According to Ahad Ha'am, it would be wrong to establish a homeland in Israel only for the sake of a body politic that lacked spiritual content.

Zionism, however, moved in the political and materialist direction advanced by Herzl. Its new person, the pioneer, was therefore to be self-realized through physical labor rather than spiritual learning.[15] Aharon David Gordon (1856–1922), a pioneer and Zionist thinker of the pre-1914 period who "came to represent that generation more than any other person" (Avineri 1981: 151), was the most famous proponent of the "religion of labor." The second wave of immigration to Israel, between 1904 and 1914, consisted mainly of young pioneers who considered manual labor central to both personal and national salvation. Gordon's views thus made him one of the most influential thinkers of the Labor Zionist movement, the Hapoel Hatzair (Young Worker) movement. The way of national rebirth, according to Gordon, "requires every one of us to fashion himself so that the diaspora Jew within him becomes a truly emancipated Jew" (Gordon 1969: 377). Physical strength, youth, agriculture, secularism, and communal socialism provided the Zionist antithesis to the degeneracy, old age, petty commerce, and religious and bourgeois life of the exile.

Pioneering did not cease to be a root metaphor even after the inception of the Israeli state. Pioneering, according to Nathan Yanai (1996), was domesticated and adopted to the new situation of state sovereignty (rather than wild frontiers) through David Ben-Gurion's concept of "good citizenship." A good citizen was a person who took root in the homeland, who subscribed to its culture and language, its creative efforts, and its vision of the future. It was a concept that was clearly embodied in the concept of *halutziyut* (pioneering). The pioneer, Yanai asserts, did not have any special rights, but voluntarily undertook special obligations; and pioneering became transient, open-ended, and dependent on the definition of the needs and goals of the social collective. Arguing along the same lines, Robert Paine (1993) claims that pioneering embodied a collective self that was accentuated at the expense of the private self.

The messianic return to nature and to a religion of labor, however, resulted more in asceticism than sexual promiscuity, contrary to Portnoy's hopes. During the second and third aliyot, between 1904 and 1938, labor was eroticized in poems, literature, and ideology, and the land of Israel depicted as a lover. Yet the gender hierarchy remained largely unchanged, and educational doctrines were puritanical. It was, in the apt words of David Biale (1992b: 182), "erotic utopianism: the sublimation of sexual desire in the service of the nation." In this manner, the Zionist ideology of sexuality was a continuation of, rather than a break from, Jewish ideology. I will return to this argument in the third chapter.

A collectivist paradigm is always interested in the formation of a "new person." This new person is identified with a new body made for him. It is often a utopian body, which must be without any individual particularities, no gender, nothing that can disrupt the homogeneity and harmony of all social components (Racault 1986). During the establishment of Fordist forms of organization in the United States, Antonio Gramsci wrote that Fordism implied "the biggest collective effort to date to create, with unprecedented speed, and with a consciousness of purpose unmatched in history, a new type of worker and a new type of man" (quoted in Harvey 1989: 126). One of Israel's greatest personifications of the pioneer myth, Yosef Trumpeldor (1880–1920), was cited as explaining the notion of "pioneering" in the following words:

> He is someone who walks ahead, . . . a much broader notion than that of [socialist] workers. We should construct a generation with neither interests nor habits, . . . a bar of metal, flexible—but metal. A metal that can be forged into whatever the national machinery needs. A wheel is needed?—I am a wheel. Screws, nails?—take me. . . . I have no face, no psychology, no emotions, no name: I am the pure idea of service. (Cited in Eilam 1973: 138)

"No face, no psychology, no emotions, no name": the embodiment of collectivism thus resulted in the erasure of individual features. Sexuality, which is an individual trait, was banished to an almost nonexistent private realm where it would not stand in the way of communal goals. The communal webs of laboring "brothers and sisters" during the second aliya were later replicated in the new combat units of the sabras (the *palmach*), where "brothers and sisters in arms" dismissed the hierarchical use of military uniforms.

The communal grip on the individual body is evident in early Israeli literature, poetry, art, cinema, and in fact any cultural domain in which the body is represented. Israeli art, for example, in the words of Meir Wizeltir,

> in its important manifestations in the 30's, 40's and even 50's, evaded the human being, denied, repressed, alienated and deceived it. . . .
> None of the serious painters working here was seriously interested in portraits or in the human figure. Humans, to the extent they appear

in these paintings, are stereotypical, a stain in the landscape. The phenomenon is amazing in its totality. It has no real connection to the conflict between realism and abstract—since it is shared here by both abstract and realistic painters. It would be even more dubious to relate it to the ancient biblical command, "Thou shalt not make any graven image," since this command did not stop Chagall, Soutine, or Modigliani. (1980: 25)

Early Israeli cinema provided yet another genre for the embodiment of collectivism. The films of the 1930's–50's always show the pioneers in groups and engaged in the same activities: working together, eating together, reading newspapers together, smoking together. This unity produced anonymity: those human figures are not individuals but prototypes, members in the "army of labor." Labor itself, however, is always filmed from afar. The abstract idea of hard work can be better shown without closing-up on sweaty faces and dirty men. Natan Gross (cited in Bursztyn 1990: 68), one of the famous photographers of that period, recalls how he was required to cut a close-up of a pair of worn-out shoes left on the floor of a workers' shack. The close-up disturbed the commissioner of the Histadrut (workers' union), who said it was inconceivable that a "new Israeli worker" should walk around with torn shoes.

The pioneers of the 1920's and 1930's gave birth to the so-called "first generation to redemption," the first generation of native-born. Occupying a privileged position in Israeli society in the prestate years, these sabras continued their halutzim parents' contempt for what were taken to be diaspora Jewish characteristics (for sociological discussions of the sabra, see Rubinstein 1977; Katriel 1986; Almog 1994; and Spiro 1965). But they also distinguished themselves in new ways.

The decade preceding and culminating in Israel's War of Independence was the sabras' formative era. The trope of the sabra was canonized through literature written during and after 1948, by what came to be known as the "1948 generation." It was in 1948 that Trumpeldor's "bar of metal" was reincarnated in the figure of the new Israeli sabra-soldier. In one of the most famous songs written during and about the 1948 war, the poet Ayin Hilel coined the metaphor of the "gray soldiers," who "fought their bloody way to a homeland, . . . reaching it with their bodies, or without" (cited in Yoffe 1989: 355). The gray soldiers were 1948's bars of metal.

One of the sabras' new characteristics was a distinctive style of communication called *dugri*, or straight talk (Katriel 1987). Netiva Ben-Yehuda (1981), herself a legendary soldier-girl of the palmach, explains in her autobiography how this style evolved. Together with a matching "simplicity" and "sincerity" of dress and behavior, it was meant to set the sabras apart from the Jews born in the diaspora. According to Jay Gonen (1971: 109), young native-born "Palestinian" Jews in Herzlia Secondary School in Tel Aviv used to challenge the star pupils (usually children of European immigrants) to peel a prickly cactus fruit called sabra. It takes a proficiency that only comes with practice to uncover the sweet insides without getting numerous tiny thorns stuck in one's fingers. The term sabra, which was first coined to express the amusement of some of the founding fathers with their untamed and uncivilized offspring, was later turned into a prickly aura of authenticity. Amos Elon, in his landmark work *The Israelis: Founders and Sons* (1971) describes the classic sabra as "the puckish figure of an eager teenager in khaki shorts and open sandals" who "exudes an air of infantile naiveté combined with childish cunning: charming innocence mixed with blunt artlessness and the bridling of untutored strength." Elon contrasts the generation of native-born sabra "sons" with the more emotional and idealistic European-born "fathers" who created the state: "Frequent and prolonged periods of service in the army breeds a stark, intensely introverted icy matter-of-factness in the young. The harsh starkness that marks sabra speech and manners stems from many years of deliberate educational efforts to produce 'normal,' 'manly,' 'free,' 'new' Jews, unsullied by the shameful weakness of exile" (ibid.). These carnal features of the sabras accentuated their "directness and wholesomeness," in contrast to the degenerate spirituality of the diaspora Jews.

The sabras had to prove to themselves and their parents that they were indeed the long-hoped-for "first generation to redemption." In *Thieves in the Night*, the novelist Arthur Koestler noted how, to their parents, the young sabras seemed like "Hebrew Tarzans." An important part of that image was concerned with military power. The Zionist socialist thinker Yitzhak Tabenkin, a pioneer and a founding father of the palmach, demanded that kibbutz teachers and nurses begin the military education of the young generation as early as kindergarten, "immunizing the child by sending him to spend his nights in the open, teach him how to use a stick and throw a stone, and make his body stronger" (cited in Ben-Eliezer 1995: 93).

Netiva Ben-Yehuda recounts how, after she expressed doubts regarding her role in a military incident where she had to kill Arabs, she was shouted at by Saul, a member of a kibbutz and one of her parents' generation:

> Fool! What will you all amount to? These are the thoughts of a weak, miserable people. Do we want a normal people here? Do we want to stop being miserable diaspora Jews? Weaklings? So among other things we have to invent the Jewish hero . . . a strong person, free, liberated, who can take a gun in his hand and kill those who want to kill him, before they do, do you hear? If you can't be like this, then you are either a sissy or a damn Diaspora Jewess! (1981: 162)

The cultural elitism of the sabras was further reinforced by their encounter with holocaust survivors. Members of the prestate community tended to resent, and reject, the Jewish victims who, according to the widely held belief, passively accepted their fate and behaved in the traditional weak manner of Jews in the diaspora (Porat 1986; Liebman & Don-Yehia 1983). This stance was combined with the socialist Zionists' ideological rejection of the diaspora, together with the ideas of militarism and state sovereignty turned into a way of life in the newly created state (Ben-Eliezer 1988, 1995). A kibbutz haggada written just after the disclosure of the dimensions of the holocaust argued that it was not only Hitler who was responsible for the death of the six millions, "but all of us—and first and foremost these six millions. Had they known that a Jew has power too, they would not have all gone as a lamb to slaughter" (cited in Reich 1972: 393).[16]

The root metaphor of "a lamb to slaughter" was not directed to the holocaust Jew alone; it was the cumulative conception of what was considered the Jews' passive acceptance of their victimization, from the time of the crusades to the Russian empire's pogroms (Segev 1991; Young 1991). Ben-Gurion, in an interview with the *New York Times* (12.18.60), justified the trial of Adolf Eichmann, head of the Gestapo's Jewish Section, on the grounds that it "prov[ed] to Israel's younger generation that Israelis are not like lambs to be taken to slaughter, but a nation capable of fighting back."

Considered against the backdrop of these various ideological strands of nationalism and militarism, it is no surprise that the cultural image of the sabra was primarily defined through the body. The palmach fighter, who became the authentic representative of the native-born, was often portrayed as "free of uniforms, one of the *Jama'a* (Arabic for 'the gang'), with short khaki pants, his blouse open to the belt, tanned face . . . in short: one of us" (from a palmach booklet cited in Ben-Eliezer 1995: 121). According to the novelist Ehud Ben-Ezer (1967: 110), the sabra was "tall, strong, tanned, simply dressed, sandals, *blorit.*" Blorit, or "high-rise hair" that was not cut or styled, became a particular cultural idiom of the "wild and untamed, yet innocent and sincere" sabra. The sabras portrayed in the literature—from Moshe Shamir's Uri (and later, Elik) to Chaim Chefer's Dudu and S. Yizhar's Gidi—all have *blorit* as their unmistakable mark (Maoz 1988). These sabra features play an important part in the screening processes that govern the Israelis' life course from pregnancy on, the subject of the next chapter.

2 CHOOSING THE BODY

PREGNANCY, BIRTH, MILITARY, WAR, AND DEATH

IN HIS BOOK about American manhood, Mark Gerzon, an American Jew, glamorizes Israel as a nation in arms that knows how to treat its fighters. He describes, for example, how as he was traveling by bus to Tel Aviv from a hospital in Jerusalem (where he was treated after a car accident) in 1969, "the women on the bus all stare at me. Whether grandmothers or teenagers, they smile warmly. Their smiles are not the stiff acknowledgment of urban strangers but the sensual, affirming, nurturing beam of old friends, old lovers. 'Why are they looking at me like that?' I ask my friend. 'They think you are a wounded soldier,' she replies" (1982: 47). With a bandage around his head, the American traveler in Israel—more than two years after the 1967 war had ended, and while Americans were still coming to terms with the Vietnam War—feels like a hero. For the two-hour ride, he is (in his own words) John Wayne. Like Philip Roth's Portnoy, Mark Gerzon finds Israel's sex appeal in the image of strong men and revering women. For these two writers, Israel is indeed the land of the chosen body.[1]

In this chapter I ask what it is like to be "chosen" in Israel, and how this process of screening and regulation actually takes place through the body. Gerzon's head bandage—the stigmata of the wounded soldier—is one of a plethora of bodily insignia that are supposed to portray manhood in the land of the chosen body. Like the *blorit* hair of the prestate palmach fighters, it stands for youthful bravery.[2] The image of the head bandage, however, implies that the chosen body also has its dark side. The first part of this chapter will focus not on the blorit, the glorification of youth in arms, but rather on its flip side, the bandage, the screening and masking of individual (deviant) bodies that do not match social expectations. In Foucault's terms, I discuss the chosen body as a discourse of normalization.

The screening of bodies and other corporal regimens are discussed here by looking at three prominent examples. These examples are presented according to their relative place in the life course. I begin with premarital and prenatal screening and continue on to parents' reactions to their newborns; then examine the screening of combat soldiers; and conclude with the screening of soldiers' corpses at the Israeli Institute of Forensic Medicine. These "cases," while drawn from very different life stages, all revolve around the chosen body and its images. They exemplify and throw into relief the patterns through which public prescriptions mold the private body. Taken together, these various incidents of testing and screening compose a generalized life cycle for the chosen body. This life cycle begins with pregnancy and delivery, continues through adolescence and youth, reaches its peak in conscription and military service, and culminates in death.

Choosing the Baby's Body

In this section, I discuss three issues: the screening that takes place before and during pregnancy, the bonding of parents with their newborns, and parents' reactions to their newborns.

Premarital and Prenatal Screening

The chosen body provides a cultural background for premarital and prenatal screening in Israel. The development of new genetic technologies has resulted in a multitude of tests, including prenatal testing (amniocentesis, CVS), individual carrier screening, and pre-implantation testing as part of in-vitro fertilization. All these tests are very popular in Israel. Although Israeli sperm banks do not screen for particular traits, they do employ other types of screening. Since 1998, one sperm bank has allowed donor insemination (DI) recipients to choose from a list of donors' profiles.[3] But many recipients are unaware of the option and use the other, secretive clinics. All sperm banks guarantee a thorough medical screening of potential donors' health. A recent study found that when asked about their preferences regarding a DI baby, men were more "loyal" to their own features than women were. For women, the preferred model was the one considered most attractive by the dominant (Ashkenazi) society (i.e., tall and white, with blue eyes and light-colored hair).[4] The authors' analysis of this observation is quite pertinent to the idea of the chosen body:

> Although presently containing different ethnic, class and national groups—to a much higher degree than that which characterized Israeli society when it started developing—Israel's political culture does not include an ethos of pluralism. Groups and sectors struggle not only for participation but also for control over the symbolic (as well as the economic and political) center. Although veteran Ashkenazi Israeli carriers of the white Western image of the *Tsabar* have lost much of their power, primordial images which were set up by this group, are apparently still hegemonic. Furthermore, the politics of appearance and beauty have to do with the "naturalization" of hegemony, in our case that of the dominant *Ashkenazi* man. (Birenbaum-Carmeli & Carmeli 2000: 28)

The preference for a masculine, Ashkenazi-like baby is also reflected in the donor population at the sperm bank that offers a choice of donor profiles (Carmeli et al. 2000). Although the ratio of Ashkenazis to Mizrachis in the general Jewish population is roughly 6:5 (Central Bureau of Statistics, 1999, Table 2.22), only three (8.8%) of the donors were of Oriental origin and another six (17.6%) were ethnically mixed. All the rest, nearly three-quarters (73.6%), were Ashkenazi. Other donor features were also somewhat biased toward a particular type of appearance. Thus, the professional staff presented a requirement for a complete high school education, though such a demand is not required by the state. Moreover, either the title "soldier" or an academic field of study appeared under the rubric "field of study" in the donors' list. None of the donors was described as a blue-collar or a nonprofessional worker.

Parents and professionals therefore collaborate in a process of screening designed to promote procreation that is geared toward the chosen body. According to another recent study, by Elly Temen (2000), Israeli couples and singles who want to use the services of a surrogate mother must undergo a long and often humiliating process of screening. They are tested by criteria of social compatibility: couples should be married and have no criminal record, and preference is given to single women who are widows or divorced.

In the case of the screening tests performed on individuals before and during pregnancy, the goal is simply to ensure the delivery of a healthy child, rather than a particular model. According to Professor Gideon Bach, head of the Human Genetics Department of Hadassah, a large hospital located in Jerusalem, Israel's world record in the use of such tests is due to the desire of Israeli parents to "assure a perfect body for their child" (personal interview, 11.7.00). The state's involvement in these tests is far-reaching. Among other things, it subsidizes all tests and IVF treatments and promotes specific programs such as testing for Tay-Sachs disease. There are also specific community-based carrier screening programs. One of the most successful is Dor Yeshorim (literally, "Straight/Righteous Generation"), developed by and implemented in the Jewish orthodox community in Israel and elsewhere. The semantics of the name should be noted: it implies that the program aims to create a generation of healthy people, meaning that those who fail to participate or to comply with the wholesome standards are nonstraight, or "convoluted." The goal of this program is to help people select mates in a community where most marriages are arranged and the termination of pregnancy is not an option (see Abeliovich et al. 1996). The program started with the detection of Tay-Sachs carriers and has been extended to three other recessive diseases. The aim is to prevent stigmatization by keeping individual test results confidential; they are stored in coded form in a computerized system to enable future matching. Members (males and females) of the community join the program at the age of seventeen or eighteen. When a marriage is considered, both sides access the program using their code numbers to check their genetic compatibility as a couple. The program has a high participation rate, mainly because it is approved and encouraged by religious leaders of the community.

Dor Yeshorim is the chosen body of the orthodox Jewish community, and it provides another example of the nondiscursive continuity between Judaism and Israeli society. It is important to point out that Israeli geneticists participate in the program as counselors and technicians, and the lab tests are carried out by the technicians of Jerusalem's Hadassah hospital. This professional participation, as noted in the Introduction, violates the ethical standards of the international community, which holds that genetic counseling should involve full and nondirective personal disclosure. Israeli geneticists may find it "easier" to accept the religious-cultural rationale behind the Dor Yeshorim program not just because of the eugenic practices common to Judaism and Israeli society (Sagi 1998), but also because of their own preferences. For although it is true that most of the carrier detection programs and genetic tests are aimed at the Jewish population of Israel, there are particular community-based screening programs for non-Jewish minorities that were initiated and are implemented by Jewish geneticists and other health professionals.

So alongside the national emphasis on procreation, which is rooted in Judaism (witness the biblical command, "Be fruitful and multiply"), there is a complementary secular emphasis on selective pregnancies. Abortion requires a special committee of social workers, religious representatives, and physicians to label the woman "a reproductive deviant" (Amir & Binyamini 1992a, b). Most legal abortions (about 80%) are the result of fetal defects. But the Israeli Law of Abortion allows the termination of pregnancy on the basis of other factors (see Amir & Navon 1989; Metzner-Licht et al. 1980; and Shenkar 1996). When young unmarried women (under seventeen) approach the committee requesting an abortion, they are usually reproached for "not using proper contraception" and warned "to take responsibility next time" (Amir & Binyamini 1992a, b). In cases of fetal defects, however, there is no moral preaching on the part of the committee, no doubt because in this case the members consider the application to be in line with the normative order and the reproductive requirements of the collectivity.

Moreover, there has never been in Israel an ethical-medical-religious discussion of the definition of "fetal defects" and why a child with a certain defect cannot be part of Israeli society. There is no public list of fetal defects that can be used to justify early or late (after the 23rd week) abortions. Until April 1999, the committee's application forms did not contain a place for describing the defect. Parents and doctors in Israel tend to go for abortion in cases that would generate hesitation in many Western countries, such as an easy case of Gaucher disease, spina bifida, or a cleft lip (Prof. Gideon Bach, personal interview, 11.7.00). For all practical purposes, eugenics appears to exist in Israel in the form of multiple tests (on a personal and community level) and a liberal approach to abortion (Prof. Dan Grauer, personal interview, 5.11.2000).

Modern Israeli eugenics represents an application of a new technology to an old Zionist idea. As Raphael Falk (1998: 594) points out in his article on Zionism and biology, Nordau's Judaism with muscles was actually a eugenic response to the Jewish condition in the diaspora.[5] Furthermore, Nordau's solution hinged on Lamarckian premises (i.e., the inheritance of acquired traits). Just as the Jews acquired in the diaspora the "effeminate submissiveness" trait, so the new Hebraic generation that had returned to Zion would reacquire the missing masculine trait. Nordau called it the "Anteus treatment," after the mythological Greek wrestler who was invincible as long as his feet touched the ground.

In Israel, the Zionists' eugenics turned into a selective prenatal policy backed by state-of-the-art genetic technology. According to Tzipi Ivry (1999), the worldview of Israeli gynecologists who are involved in legitimate abortions is based on military terminology. Physicians who regard themselves as "commando fighters" will justify abortions by the military thinking that sees killing as a necessary means for attaining its goals. In most Western countries, late abortion is legally prohibited except in medically necessary circumstances. In Israel, the law permits late abortions and is used by many parents to lodge malpractice suits against hospitals that allegedly failed to warn them about fetal defects, especially we may assume if the Israeli Ministry of Health issues a complaint against them, as it may do, for allowing the birth of a defective child that now costs the state a lot of money.

Conditions of Bonding: Parents and Their Newborns

During the years 1985–91, I studied, as a Ph.D. project, parents' reactions toward their appearance-impaired children.[6] What caught my attention was a pattern of rejection—either by leaving the child in the hospital or by abusing the child in various ways at home. About 50 percent of the impaired children were rejected altogether. About 80 percent of the impaired children who were taken home ("adopted") were kept secluded there (Weiss 1994). Israel is number one in the world in its rate of child rejection due to impairments.[7] My observations indicated several external attributes that, once spotted, could result in the child's rejection by the parents. Among the most notable were bluish skin color, an opening along the spine, a cleft lip, a lack of proportion in the facial features, and medical apparatus attached to the body or openings made in the body. Moreover, I found that parents were bothered more by external, openly visible impairments than by internal or

TABLE 1

Parental Reaction to Newborns by Type of Defect

(N=350)

Parental reaction	Internal defect (no visible signs)		External defect		Total
	Number	Percent	Number	Percent	
Abandonment	7	7%	171	68%	178
"Adoption"[a]	93	93	79	32	172
TOTAL	100		250		350

SOURCE: Weiss 1994.
NOTE: The percentages are rounded up.
[a] Indicates various types of abuses in the home.

disguised defects. The extreme significance of appearance in the case of the newborn is illustrated in Table 1, which summarizes the findings of the study.

Although openly condemned in Western societies, the rejecting of children with defects is frequent in Israel. Sadly enough, such practices are part of human history in general.[8] In theory, this state of affairs was supposed to have been corrected by the advent of technological progress and the welfare society. But even in contemporary "enlightened" Western society, we find that the stigmatization of impaired appearance often persists, although infanticide and mortal neglect have tended to be replaced by hospitalization and social denigration. A. H. Beuf (1990), in her study of American children, found that such mainly aesthetic disorders as psoriasis, cleft palate, obesity, and myopia are often deeply stigmatized.[9] The stigma thus persists in spite of economic and technological progress. However, although the rejection of appearance-impaired people is a worldwide phenomenon, Israel is the world leader in the proportion of children who are rejected for physical flaws.

In accounting for child rejection, anthropologists have traditionally invoked either religion (usually in the case of "simple" African societies) or material hardships (as in third-world countries such as Brazil). In modern Israel, these two factors do not play a major role, and child rejection must therefore be linked to other cultural scripts. I relate the stigmatization of impaired appearance to the culture of the chosen body after presenting the following examples. Taken out of the overall data, these cases are representative of the entire study population that I observed between 1985 and 1991.

"Is Everything in Place?"—Mothers' First Reactions
to Their Newborns

I begin with a description of the behavior exhibited by Israeli mothers immediately on giving birth to a normal child. Their remarks tended to follow a pattern. The moment the child was born, the mother expressed interest in the newborn's sex. At this point, some mothers asked to see their babies' sex organs in order to confirm the midwife's proclamation of the child's sex, a proclamation that generally elicited either a response of delight or an expression of disappointment. This observation echoes G. P. Stone's (1962) argument that sex is indeed the primary component of external appearance in establishing relations.

The next reaction was the same regardless of the newborn's sex. At this stage, when a connection between external form and normalcy was already established, mothers began to examine their babies visually. They checked his or her body limb by limb and made comments on the coloring, size, weight, number, cleanliness, perfection, and proportion. Consider the following remarks of new mothers from different ethnic and educational backgrounds: "Is he O.K.?," "What a cute girl! She's really big . . . and pink," "Is he all right? Not missing anything?," "Everything in place?," "Why are her fingers so small and blue?" "Why is his head so big? Is that normal?," "Why is she so bald?"

If the baby passes muster in this external examination, it is called "sweetie" and other names of endearment, and its acceptance into the family is indicated by references to some external similarity between it and another member of the family. However, if the newborn does not pass the test of appearance, or worse, shows evidence of some impairment, patterns of rejection might come into play.

The child with the tail. The Meshulams' son was born with spina bifida. Mrs. Meshulam, a twenty-one-year-old native Israeli of Yemenite extraction, was a housewife with 10 years of schooling. She and her husband identified themselves as religious and had another son. The midwife noticed the serious defect the moment the child was born, since part of the membranes of the spinal cord was exposed and protruding.

Mother: "Wonderful, is it a boy or a girl?" Midwife: "A boy." Mother: "Oh, I'm so happy. I so much wanted a boy. Let me see him." Another midwife: "Yes, but he seems to have something on his spine." Mother: "Is it serious? Is it crooked? Are they going to operate on him in order to remove the crookedness?"

The doctor, who was summoned immediately, explained the situation to the parents, saying that "the child has no chance of surviving for more than a few months. We cannot repair such a defect. My advice is that you sever contact with the child." Heavyhearted, the parents remained silent. Then they whispered to each other and finally came to a decision. Mother (emphatically): "We want to see the child." Doctor: "I don't think you should. It will be hard for you." Father: "We insist."

The nurse brought in the child, who was already dressed and looked very cute. Mother: "What a doll! You look so cute." Father: "Really beautiful." Mother: "He is such a lovely child. Your saying that he has a defect and is going to die just doesn't make any sense. I certainly won't cut off contact with him." Father: "Exactly. We're going to take him home." Nurse: "O.K., give him to me now so I can diaper him." Mother (distrustful, suspicious that he will be taken away from her): "First we want to see just what it is our son has on his back." Nurse: "Maybe tomorrow?" Mother: "No, right now."

The mother kissed and hugged the baby, then grudgingly handed him over to the nurse. The nurse undressed him and showed the parents his deformity. The parents immediately recoiled as if they had received an electric shock. Mother: "Get him out of here. Take him away" (motioning away with her hand). "What an ugly thing!"

For two days, Mrs. Meshulam did not go to see her baby. She told a ward nurse whom she had befriended that she had given birth to a "baby with a tail," and that she was not going to let "such an ugly thing" into her house. When she was discharged from the hospital, she left her son there, and she never returned to see him.

Like other mothers I observed, the first thing this woman was interested in was the sex of her child. She was happy to hear that she had a boy. The common greeting, uttered by midwives and nurses in such cases, is, "Hopefully, he will grow up to become a fine soldier." Next, the mother's interest shifted to his external appearance. She cast a rapid, superficial glance at her child (because he was speedily whisked out of the room), enough to see his body and his hands and feet. The parents' momentary decision not to sever contact with the child reflected their perception that the child's external appearance was more important than "what the doctor said." But the rapid reversal in their decision was founded on the exact same perception; they were repulsed by the child the moment they saw his deformity as it was, bare and undisguised.

Two cases of cosmetic repair. Gabi suffered from spina bifida, but in this instance, the defect was very slight and curable. In a manner parallel to other studies of appearance-impaired children who underwent cosmetic surgery (see Beuf 1990), this case serves to illustrate the change in attitude toward the child before and after the operation. On first encountering Gabi and his physical deformity, the parents—secular Jews of European extraction—reacted with complete rejection and disclaimed any responsibility for him, saying "He is dead for us." This reaction was not altered by the favorable prognosis given by a senior physician. Only after the external deformity was removed did the parents consider "readopting" their child—not, however, before cautiously examining his looks so as to assure themselves of the repair's success.

In this full transformation from abandonment to acceptance, the "cosmetic repair" operated as a rite of passage, involving even the ceremonial act of renaming: as the father announced after the operation, "We have decided to change his name to Samuel. If a new son has been born to us, he should have a new name, as well."

Miriam, another child born with spina bifida, was abandoned at birth. She had an operation that removed the physical deformity, but left her with a very poor prognosis for recovery. The mother, informed that the physical deformity was removed, arrived at the hospital, after months of neglect, checked her daughter's naked body, and thereafter remained at her bedside and nursed her. Unlike Gabi, whose cosmetic repair removed all the symptoms, Miriam was left paralyzed. Yet the similar change of heart demonstrated by the parents attests to the overarching significance they attached to the external deformity.

This preoccupation with appearance is borne out by the many "normal" children in my study born with a cleft lip or lack of proportion in their facial features who were abandoned at the hospital even though the prognosis for cure was excellent. It seems clear that the rejection by parents was due not to the severity of the disease but to the visibility of the deformity. In addition, the acceptance of formerly appearance-impaired children was seen to take place after the deformity had been removed, irrespective of the severity of the disease as established by the doctors.

The monster's ghetto. This case is different from the others in that it deals with an appearance-impaired child who was taken home by her parents (if not at all willingly). It illustrates that although parents may consent to take the child home, rejection can still take place within that supposedly safe environment.

Pazit's parents resided in a town in the central region of Israel. They were Moroccan Jews, thirty years old, who had been brought to Israel as infants. They had only an elementary education. The mother was a housewife, considered to be warm and caring toward both her healthy sons, ages five and six. The family resided in a two-room apartment.

Pazit was born with external and internal problems: asymmetry of the facial organs and defects of the heart and kidneys. Her chromosome formation was normal, and there was no indication of mental retardation. Two days after delivery, Pazit's parents requested that she be transferred to an institution. They maintained this position even when, after several weeks, all tests confirmed that the baby was mentally normal. The mother explained: "It's difficult for us. We are good parents, but this girl we do not want at home because she's sick and looks like a monster. She is blue [due to a heart defect], each ear is different, she has a large nose. Everyone who sees her is appalled." The mother burst out crying. "It's impossible to accept her condition and her appearance. We don't want this girl." The mother concluded: "We can't bring this girl home. I'm willing to visit her at the hospital, but not to take her home. Where will we put her? We have a small home. We can't isolate her so that no one will see her. No one will want her in his room. We don't have a balcony to put her on, and if we put her in the corridor, everyone will have to see her. We are not taking her home."

The parents' refusal to take Pazit home persisted for a few months. Over this period, they shut themselves up in their apartment, drew the shutters, and refused to admit strangers, fearing that Pazit might be unexpectedly brought there. Once a week, the mother visited Pazit, accompanied by the local social worker.

When Pazit was eight months old, the hospital authorities filed a complaint of child desertion with the police and demanded their intervention. Two days later, Pazit's parents received a cable informing them that the girl would be brought home the following day. Early the next morning, a hospital nurse, accompanied by a policeman, brought Pazit to her parents' apartment. When their ringing at the door remained unanswered, the policeman put Pazit, well wrapped, at the door. About two minutes later, the door opened, and Pazit was hastily taken in. I had arrived at their home about an hour earlier and observed the parents' behavior.

Very silently, the father picked Pazit up and said, "Where will we put her?" Mother: "In the living room?" Father: "That's a problem. How will we watch television?" Mother: "Maybe we should put her in the kitchen?" Father: "Impossible. We eat there." Mother: "So, we'll put her in the corridor. There is no other way. The children will play in their room or in the living room."

And so, Pazit was put in the corridor. The father removed the light bulb, leaving the corridor in total darkness, so that Pazit could not be seen. "It's a ghetto for monsters," said the parents. The mother tended to Pazit reasonably well. She was kept clean and had no wounds. Yet, about every two weeks she was hospitalized. During these periods, the corridor was lit again, and the house was opened to visitors. "We have to breathe once in a while," said the father. The mother also "tried to rest" when Pazit was in the hospital and preferred not to visit her. Pazit passed away when she was one year and three days old.[10]

My interest in appearance-impaired children reflects an interest in extreme human situations, situations that test our basic cultural scripts—parental bonding, in this case. What drives the parents I "studied," what drives me as a parent. How would I behave in their place. Looking at these parents, I felt shamefully happy that my children were healthy.

Here, at the hospital, life was so intense. Life becomes intense when it is on the borderline, on the edge. Nothing else is relevant. It isn't relevant if my son got a bad grade in class today. All that is important is happening here and now. I think this is why TV hospital shows are so popular. They capitalize on our vulnerabilities, seduce us into tragedies that are bigger than everyday life.

One day I sat in the hospital with one of the doctors, a tall, beautiful young woman, but also pale and "fading up." She told me, "Let me show you the room that we hide from everyone." And she took me into a room that was in between two sections. It had little children whose tiny bodies and faces were convoluted. She told me that the parents of these babies had all disappeared, abandoned them. The hospital had issued an instruction to move the babies out. "But I know," the doctor told me; "I know that if they would be moved out, they would be lost." Lost, she is telling me. There were no tears in her beautiful eyes as she was saying this. "Tough guy," they call her. She decided to keep the babies in that room, where only she and another nurse took care of them. I asked her, then, "What's happening with you?" "I'm okay," she said. "I don't have time but I'm okay." "No," I repeated, "how do you manage being there, in that room?" She looked at me in surprise. "I don't know," she said. "It's like being in an upside-down world."

Is it like being in an upside-down world? I am now asking myself. I don't know. I've been caught in this upside-down world for so long, it is the only world I know.

Body Image and Parenthood

Although difficult to accept, the behavior just described represents a common pattern. As Table 1 shows, most of the children suffering external defects (68.4%) were abandoned by their families, even though most of them did not suffer life-threatening illnesses and in certain cases the defect was only severe aesthetically (e.g. a cleft palate). In contrast, most of the children suffering internal diseases were not abandoned, even in cases of serious illness where the chance of recovery was slim. These findings cast profound doubts on the "natural" and "regular" process of parent-child attachment called "bonding." Nancy Scheper-Hughes (1991), who found similar types of child neglect among Brazilian mothers, relates it to the "struggle of survival" taking place under conditions of high mortality and high fertility in poor urban suburbs. Since this kind of materialist explanation is not relevant to the conditions of life in Israel, let me suggest other possible explanations.

First, are we perhaps dealing here with an ethnically based, specific sociocultural phenomenon, whose causes are to be located in the tradition-al culture of Israeli immigrants? Indeed, other anthropological studies have explored parts of Israeli society as traditional social groups undergoing accelerated processes of modernization; when confronted with modern medical technology, members of these groups reacted in apprehension and refusal (for a discussion of the case of impotence, see Shokeid & Deshen 1974: 151–72). "Bonding," then, must have, as Elizabeth Badinter (1981) and Scheper-Hughes (1991) each argued, "sufficient conditions" (namely, the bourgeois nuclear family) in order to be successfully employed as a repro-ductive strategy.

Notwithstanding the possible validity of this proposition in other instances, such an ethnically based cultural explanation would have to be ruled out for the Israeli situation, since the observed phenomenon cut across ethnic, economical, and educational categories. The overwhelming frequency of the observed phenomenon could be readily explained according to sociobiology as an evolutionary mechanism for the survival of the fittest. However, this general explanation does not account for the rejection of newborns with only slight external deformities.

An alternative explanation is that the chosen body is an imported image from North America, copied from the fashion of body-building, plastic surgeries, and the fitness craze. But in the United States, there is a smaller number of parents who are willing to reject their child on the basis of external defects. On the other hand, the United States also instituted many federal laws that protect people with disabilities, laws that do not yet exist in Israel. In Israeli society, the disabled body is almost totally excluded. It becomes a non-person in a way that enables its rejection, institutionalization, and seclusion. The pattern of rejection that I found therefore seems to emanate from a cultural source that is unique to Israeli society. My interpretation sees this source in the cultural script of the chosen body that has emanated from militarization and chauvinism. Following Mary Douglas (1973), one may hypothesize that a society deeply concerned with external borders is also deeply concerned with body boundaries.

Israeli society provides a direct correlation between the construction of parenthood (especially mothering) and the reproduction of the nation. As Susan Kahn (1997: 82) argues, "When the dominant religious culture provides the conceptual groundwork for kinship, as it does in Israel, and when this same religious culture determines identity matrilineally, as Judaism does, then eggs and wombs are not only the variables that determine maternal and religious identity, they are the variables that determine citizenship as well" (see also Kahn 2000).

Kahn's important study on new reproductive technologies in Israel (such as artificial insemination) provides empirical findings that strengthen my study. Although she is interested mainly in the relationship between halacha and modern medicine, her study inadvertently but succinctly shows the impact of the nationalist ethos on reproduction. "If you're not a mother, you don't exist in Israeli society," a social worker in a Jerusalem fertility clinic told her (Kahn 1997: 107). And when Kahn analyzes the procreation stories of several unmarried Israeli women who have chosen to become mothers via artificial insemination, those stories provide a unique perspective on state support for the assisted conception of children by *unmarried* women (which goes against Jewish religious conduct). It is an unprecedented phenomenon, in halachic terms, that is promoted by the Israeli nation-state.

The disturbing findings on parents' rejection of their appearance-impaired children led me to a seductive yet elusive concept: body image. Body image is uniquely significant in the very first stages of bonding, since no other information concerning the newborn really exists. The image of the body is therefore regarded as being also, one might say, the image of the "soul." This assertion is strengthened by many traditional beliefs that connect an appearance-impaired body to some sin performed by the person (possibly in a previous life). It is this extreme significance of appearance that accounts for the abandonment of more than 68 percent of the appearance-impaired newborns, whereas the 93 percent suffering from internal defects were "adopted." The congenitally deformed infant challenges the tentative and fragile symbolic boundaries between human and nonhuman, natural and supernatural, normal and abominable. Such infants may fall out of category, and they can be viewed with caution or with revulsion as a source of pollution, disorder, and danger. This can result in the stigmatization of the appearance-impaired child as a "non-person," and lead to his or her rejection.

But what actually is this body image, and where does it come from? Although there is a biological answer to this question, as an anthropologist, I am more interested in the role that culture plays in reproducing certain "internal representations" under specific social circumstances. Beauty, obviously, is based on cultural ideals. My question is, therefore, what kind of social circumstances enabled—and what sort of normative control produced—the body image inculcated by Israeli national ideology, the image that Israelis have in mind when they screen their children as future citizens (good soldiers and wives). My search for an empirical grounding of that oblique notion provided the impetus for a 10-year study of drawings of the body made by university students, which will be described later in the chapter. Before proceeding to the generalized Israeli conception of body image, let us examine the consecutive stage in the life course of the chosen body: conscription and military combat service.

Choosing the Combatant's Body: Conscription and Military Service

Conscription and military service are the ultimate social goals of the chosen body. This is arguably the reason for the screening of the newborns of the previous section. Newborns are preselected to fit the mold of the chosen body; eighteen-year-olds are selected to become the chosen ones. In Israel, with very few exceptions, every (Jewish) citizen who reaches the age of eighteen is recruited into the IDF (Israel Defense Forces). Israeli law exempts only the physically handicapped, the psychologically incompatible, ultra-orthodox Jews, and married women. Girls can be exempted from military service by declaring that enlistment is contrary to the religious principles of their families. They may do a one- or two-year period of national service instead. The duration of army service is three years for men, and two (or two and a half) for unmarried women. After discharge from this compulsory military service, men (and some women) serve in the reserves for 30 days a year until the age of forty-nine. Officers may be called upon for additional days.

Recruits are tested and selected in four general areas—education, intelligence, Hebrew, and physical condition. Applicants for elite combat units are tested during the year before conscription. The successful ones then go off to those units for special, "in-house" training, while the majority of recruits undergo six months' basic training in military camps. The legitimacy of military screening is largely due to its perceived role not only for survival but also as educator of national values. Military indoctrination begins at school: the IDF and the Israeli school system cooperate in the form of a paramilitary class. Formerly called Gadna (Youth Corps), its name was changed some years ago to Shelach (a combination of *sade,* field, *leom,* nation, and *chevra,* society). These state-run activities socialize Israeli youth in the ways of the army, including the use of rifles, map-reading, and calisthenics (for more details, see Lomsky-Feder & Ben-Ari 1999; and Levy-Schreiber & Ben-Ari 1998). Various representatives of army units are invited to the schools, and "school excursions" to military bases around the country are common. The aims of these diverse activities are to prepare young Jewish Israelis both physically and emotionally for army service. Combat units are given high visibility and prestige in these preparations (see Israelashvili 1992). This hierarchy of combat versus noncombat roles sustains the ethnic hierarchy that pervades Israeli society. In 1980, for example, Gadna leadership was found to come for the most part from middle-class and well-established families, only one leader in five being of Sephardic origin (Roumani 1980: 83). Alongside the military's vocational involvement in high schools, it also educates those of its recruits who have little schooling; besides the usual professional training, they are given courses in Zionism and Hebrew.

The IDF is publicly seen as a significant mediator among the components of the Israeli social system. It provides a meeting ground for different ethnic groups and facilitates a fusion between the personal body and the body of the land (*guf ha'uma*). Even today, the IDF is still considered a model of the melting pot doctrine that aims to mold different socioeconomic groups into a homogenous national society (Lomsky-Feder & Ben-Ari 1999; S. Cohen 1997). The IDF, for example, requires combat troops to acquaint themselves with the land of Israel "through the belly"—to "[walk] over every corner of Israel, lie in ambush in every valley, sit for days in lookout posts and learn every bush and every clod of soil in the landscape" (Roumani 1980: 99). It has a special education division responsible for many special "land courses," as well as courses on the Jewish and Israeli heritage, which usually deal with such subjects as the redemption and unity of the people of Israel. The study of the land pursued in the IDF is virtually nonexistent in other armies.

Going through my high school diaries, I find many entries that relate to Gadna activities. I describe these trips, field days, and volunteer work with a cheerful, committed tone. Only now do I realize the dual tone of these descriptions. I can remember very sharply how I hated everything that had to do with physical effort: the ritual of climbing on mountains during field days, and the humiliation I suffered because I could not walk and was transported by bus. Despite all my resentment, my diary—a personal text meant for my eyes only—reflected different feelings. Even my personal writings were written under the gaze of others. "For such a landscape it is worth it to suffer," reads an entry I wrote when I was fourteen years old. But also, three years later, I dropped out of the youth movement in which I was active until then, the Scouts. This must have been a crucial decision. Dropping out of the youth movement was worse than dropping out of school. It was perceived as aban-

doning one's ideals, one's commitment to the nation. The reason for my deci-
sion, I wrote in my diary, was that the Scouts did not permit a member to wear
nylon pantyhose (which seemed too self-indulgent and bourgeois).

I described in the Introduction the connection between Judaism with
muscles (or the masculinization of Judaism) and the return to the land of
Israel, as preached by Zionist malestream thinkers. Obviously, the IDF had
a major role in the socialization of the new Hebraic man. If diaspora was the
disease, the land was the cure, and the military the necessary means of
achieving that cure. The IDF was therefore empowered, from its very inception, with educational functions. The military was regarded as a major
avenue for the absorption of immigration and as the major equalizer of ethnic differences. On the one hand, the army is the culmination of an all-
Israeli initiation course that begins in the youth movement, continues in the
Shelach, and ends in military service. On the other hand, the IDF is at the
same time a point of departure into adult Israeli society, a springboard into
professional and political careers. Myron Aronoff (1989: 132) claims that the
"primary rite of passage that initiates one into full membership in the
Zionist civil religion is service in the IDF. It is the single-most-important
test, particularly for males, for individual and group acceptance in the
mainstream of Israeli society." We might gain a better insight of the structure, elements, and functions of this rite of passage if we bear in mind that
it is a passage to both civilian life and manhood. As Ben-Ari (1998: 112,113)
contends, "The combat schema is also a schema for achieving and reaffirming manhood. Being able to act as a soldier in battles encapsulates the
notion of a man mastering a stressful situation, and, if successful, of passing
the test."[11]

In this section, I study the screening of bodies for conscription and combat service close up. My empirical data are based on 200 interviews with
parents of combat soldiers. The interviews were conducted during 1994–95
and 1997–98. They lasted from one to three hours, were usually carried out
in the parents' home, and were usually taped. All the parents were Israeli-
born and were of Sephardi as well as Ashkenzi background. All respondents
had at least one child in the military during the interview or before. The
interviews followed an open and informal procedure, guided by the request,
"Please tell me about your son and the military."

The main finding was that an overwhelming majority of the parents (193 of 200) affirmed and justified their son's decision to volunteer into a combat unit. The majority of parents (120) had actively encouraged such a decision. Seventy-three said that though they did not openly encourage that decision, they did not oppose it, and did what they could to facilitate their son's welfare during training. A small minority (7) had qualms about their sons' choice of combat duty and would rather they had chosen a more professional-academic career within the military. That is, their objection was not to military service as such, but to the risk and difficulties inherent in combat units. Five of the seven were bereaved families.

The findings thus present the Israeli family as a major agent of socialization and normative control. This role was traditionally dedicated to the family through religion, and later through religion's modern successor, nationalism. As George Mosse (1996: 19) claims, "The family gave support from below to that respectability which the nation attempted to enforce from above." Such respectability included hierarchy and order, sex role socialization, and the policing of sexuality. The family was supposed to mirror state and society; "through the rule of the father as patriarch, the family educated its members to respect authority" (ibid., p. 20) and helped reproduce national ideals of virility for men and domesticity for women. Friedrich Ludwig Jahn, the founder of the Gymnasts and the fraternity movement—both German nationalist revolutionary movements—had called the family "the fountainhead of the national spirit" (cited ibid.).

For an overwhelming majority of my respondents, who represented an ethnically heterogeneous, non-orthodox Jewish population, military service was a matter of consent. Parents spoke in these interviews with the affirming and standard voice of conscripted agents of the body politic. When the soldier-son "made it" into a top military unit, his success was a great source of pride for his parents. When he failed to reach a desired target, they often imputed his lack of success to minor physical problems. In doing this, parents' talk of their son's body carried the social gaze into the personal realm. They often described at length the body malfunctions that failed their son. In highlighting these bodily measures of screening, they participated in the discourse of the chosen body.

One of the mothers told me, for example: "There were moments when my son, Roy, came to me saying that he could not take it any more [the training course]. I told him, 'No, you can't give up, you're almost at the end of the course, you have almost made it.' And he told me, 'Mom, you should know it's only in your head. It doesn't depend on your physical strength as much as it depends on your willpower. If you really make up your mind about it, there's nothing you can't do. I said, 'Yes, but the ordeals you have to go through.'" She went on to explain to me: "They are like animals, throw obstacles on you, put you in freezing water in the middle of the night, in the winter, in the rain, just to test your limits. And you know, it was only because of his eyes that he didn't make it. He was operated on when he was two years old, in order to take care of a minor myopia. It was all taken care of, but the medical files don't lie. He passed all the tests except for this one imperfection."

The list of problems that parents gave as excuses for their son's failure was quite varied and not limited to physical flaws only. In addition to short-sightedness, fatigue, childhood asthma, or the occasional (yet critical) fracture/wound, they cited social difficulties, the lack of the proper "strings" that had be pulled, and even a onetime psychological breakdown. By emphasizing the "one imperfection" that failed their son, parents actually succumbed to and reinforced the discourse of screening that underlies the culture of the chosen body. The results of military screening—before and during conscription—were assumed to reflect an objective and neutral assessment. Parents never raised the possibility that military testing could in some way be culturally biased (in a similar manner to IQ tests, for example) or in any way tainted. The military decision was consequently taken for granted as final and beyond refute. Soldiers as well as parents took the role of docile subjects of a far-reaching normalizing discourse that works through the body and is dominated by the military. I argue that conceiving military testing as a passage to manhood is seen as justification for the long and painful ordeal of military training. Completing basic training and/or specialist courses, taking up new tasks and missions, entering a new unit or undertaking a new role—all of these involve stringent scrutiny and testing by superiors and peers. In terms of manhood, this never-ending ordeal fits the cultural construction of gender not as an inherent state but rather as a continuous performance (Butler 1990; Gilmore 1990). The passage to manhood, as embodied in the military, is performed mainly though the body.

I recall my own experience of military service. The unconditional support of my parents, which was—in hindsight—actually conditional upon my military achievements, geared toward keeping me on the right track, encouraging me to pass the tests of military screening, from basic training to the officers' course. (And, once more, my passivity and obedience were only occasionally hampered by bodily "resistance"—a leg "case," according to Mom, but actually I didn't have anything. My body resisted.) For example, my mother wrote, "Don't let them see how difficult it is for you. The worst is when they see your tears . . . Eat and drink and then you could run too." My parents' letters encouraged me not to give up, to be stronger, to finish what I started. The tragic irony is that my husband's letters to my daughter, Tami, during her officer's course are so similar to my parents' letters to me.

The Saga of the Chosen Body: Screening as Fantasy, Humiliation as Belonging

Whatever the final outcome of the screening, parents depicted it as extremely difficult and demanding. Once again, this depiction did not involve any criticism, but rather adhered to the glorifying view of the chosen body of the fighter. The construction of the saga of the selected body usually begins a few years before actual conscription.

Yuval's mother: "With my second son, Yuval, there was a problem, he was so skinny, we were sure they were not going to conscript him at all. When he was seventeen, his peers began speaking about conscription, units to volunteer to in the military, and so on. It became the talk of his class. He was not exactly in shape. He decided to join a physical training group. I was really surprised how he started training seriously. He even took one of the first places in a running competition that was held in his senior year at high school. He hates it when I say it—but he really took to sports. By the end of the year he had put on five more kilos."

It is important to point out that what parents told me in the interviews was actually rather similar to what I was used to hearing as a native-born Israeli, from friends and relatives and parents of combat soldiers. These stories represented a folk genre of "test stories" or "manhood stories" related to the ordeals of the military service. Edna Lomsky-Feder (1994, 1998) has collected an impressive corpus of life stories of veteran soldiers that exhibit the same characteristics of the genre I delineate here. Lomsky-Feder notes that these stories have a significant "socializing" value, since young people easily identify with the heroic attributes of veterans: grit, unyielding adherence to goals, ability to withstand physical and emotional pressure, and loyalty to friends.

Similar stories can be subsumed under the "preparation" paving the way for the "real'" ordeal, which is the training course administered by top units for selecting their recruits. Access to these screening courses is limited, and volunteers are admitted on the basis of pre-screening. Stories about this utmost test were abundant. Four statements by the mothers of sons who underwent the process will illustrate the point. One told me, "First he [her son] underwent medical and physical examinations, then he was tested for the air force. In the interviews the army did with him, he said his first priority was to be a pilot. If not that—a commando. You see, he was always the first in everything he did. . . . He didn't tell me much [about the screening course]. They had to move heavy rocks to the top of sand hills. At night too. It sure was difficult, but he never complained to me." Another said, "The training was in some faraway place [in Hebrew, *chor*, literally a hole, a word that kept appearing in the stories, also in the more idiomatic form *chor shechua'ch el*—a godforsaken place]. They slept in tents, in the desert, no food, no hygiene . . . Running from dawn to sunset, conquering mock targets, sitting in night ambushes. He did this after not passing the screening for pilots, but he told me, pilots are pussies with white hands, the real fighters are the infantry." Still another mother stated: "We were so proud that Tom underwent screening. He trained at home before that, he used to run, and I told him, you should be running with a heavy backpack, for the practice. . . . He didn't make it in the end, but not because he wasn't good, simply because they had limited room, only for the very best." The following statement seems to me especially revealing: "The training was so tough, treks in the desert, water drills, jumping into the cold sea in the middle of the night. It was because they were selecting the best of the best. I knew that my son had to try it; it was the ultimate achievement of masculinity for him."

Stories of ultimate bodily ordeals kept appearing in detail: running, climbing, trekking, conquering, a nonstop merry-go-round of physical mortification and hazing. The majority of parents and combat soldiers, however, lacked the critical perspective to see any of this as hazing. It was part of training and clearly justified as such. Klaus Theweleit describes a similar process in the approach of the German military in former times. It is perhaps a universal trait of all armies, this molding of a young boy into a soldier by reconstructing his body. In the German training camp, he notes:

Everything is planned and everything is public. . . . Any boy found hiding his head under the pillows is labeled a sissy. Boys who want to go to the toilet at night have to wake the duty officer. Cadets are often deprived of food, leave, or the opportunities for relaxation. In cadets who wish to remain such, all this very soon produces a quite extraordinarily thick skin, . . . which should not be understood metaphorically. (Theweleit 1989: 14)

In Theweleit's description of the biographies of 250 Freikorps warriors in the 1920's, a prominent place was given to the "reaching of the ultimate limits" through body exercise. Although the historical and ideological conditions of the Freikorps and the IDF are very different, the underlying culture of masculinization is similar (on the universal context of military socialization, see Arkin & Dubrofsky 1978). Interestingly, Theweleit (1989: 108) also links the body formation in the military to "the body of the people (*der Volkskorper*), to which in some strange way men who require an external 'extended self' to achieve 'wholeness' feel bound. For Freud, the origin of these mysteries was to be traced to the bodies of the parents."

The body is therefore mortified not only in order to be selected, but also in order to belong. The body mediates both the collective belonging of the individual and his elite status. It is through corporal ordeals that sons and fathers, as well as soldiers and the body politic, "become one." Yet activities that are presented as part and parcel of the screening process are often acts of hazing in practice. In 1995, the Israeli media publicized the case of a cadet in an elite navy unit who had been covered in a mixture of urine, excrement, gun oil, flour, and sand and then kicked by his mates to mark his promotion. The cadet denied that he was humiliated by the ceremony. He also denied being physically abused, and that excrement and urine were used. But photos given to Israel TV's Channel 1 by a bystander, who had since completed his military service (see *Jerusalem Post*, 4.26.95, p. 12), bore out the news outlets' reports. This sailor's story was not unique; other combat soldiers in different units all describe such ceremonies as commonplace and say they were performed with their consent. In the IDF's School for Anti-Terrorist Military Proficiency, there is a routine called "aggression training," during which soldiers have to endure constant battering without defending themselves. The declared goal is to increase one's threshold of pain. The practical outcome is that many soldiers drop out of this course, some with life-long disabilities (*Yediot Acharonot* daily newspaper, 5.7.99, p. 15). They are required to undergo such drills as "walking on soldiers" (in which soldiers lie on their backs in a row, and the trainer or a fellow soldier runs across their bellies), the "death roll" (in which a soldier is clubbed as he runs the gauntlet), the "throat fall" (in which a soldier runs toward another soldier, who moves aside at the last minute and tries to bring him down by throwing a straightarm at his throat), and the "narrow path" (in which two groups of soldiers run into each other in a narrow passage).

In 1994, the annual IDF ombudsman's report particularly attacked the problems of hazing and relations between officers and new inductees. The ombudsman, Brigadier-General (res.) Aharon Doron, wrote in the report that officers did not allow "negative acts" out of maliciousness; it was simply that they misinterpreted the aims and means of instructing junior soldiers, either from ignorance or from inexperience. The report was based on some 8,600 written complaints that reached the ombudsman's office, a slight drop from 1993. The perennial complaints about insufficient sleep, the report claimed, showed that officers did not abide by regulations, and the problem had become a norm. Complaints showed that hazing was commonly practiced at three of the IDF's six regular training bases.

Ceremonies of bodily humiliation do not begin in the IDF, but are rather an informal part of Israeli youth socialization, or at least in extreme rightwing circles. According to the *Jerusalem Post* (10.6.94, p. 3), twelfth-graders at the Pirhei Aharon (Flowers of Aaron) yeshiva in Kiryat Shmuel, which is affiliated with Bnei Akiva,[12] brutalized a group of 30 ninth-graders during an initiation ceremony. They marched them blindfolded through the main streets of the town and later pushed their faces into the sand on the beach. Boys who cried and begged for the ceremony to stop were punched and kicked by the youth movement leaders, the report said. Some of the boys were thrown from stretchers into the sea, and others had eggs broken over their heads while they were lying half-naked and facedown in the sand, digging holes. In the final stage of the ceremony, which was probably fashioned after the initiation rites (*zubur*) of the IDF, the boys had to march through a two-kilometer stretch of sandy beach, blindfolded and with their hands on their heads, all the while shouting, "We have been humiliated!" Then they read the "ten commandments," and, kneeling, had to recite: "Thou shalt not curse a twelfth-grader on penalty of death." At the end of the ceremony, one of the youth movement leaders, identified as Yeshaya, reportedly told the boys: "You can now join Gush Emunin,[13] and you will get permission to kill Arabs." A leader of the Bnei Akiva yeshivot, Rabbi Moshe Zvi Neria, said that he had repeatedly expressed vehement opposition to such ceremonies and tried to stop them, thus revealing that the reported hazing was far from an exception.

I have chosen to describe here several hazing rituals, in the army and the youth movement, because hazing is an extreme example that illustrates the general process of screening through the body. Such hazing rituals are common in the army. Their generic name, *zubur*, is an Arabic word denoting the penis. This is no coincidence, for many of these rituals contain a strong homoerotic content. Liora Sion tells the story of a captain in the parachuters who subjected new staff officers to a unique rite of passage. As one recounted:

> Among the staff members this rite was known as "the stroke." While talking and laughing, . . . I was dragged by other soldiers into the captain's tent. It seemed like a part of a very physical game that was not planned. But once I was inside the tent, everyone grabbed at me and tried to undress me. I put up a fight, but they succeeded in pulling down my pants, exposing my behind. Then the captain stepped forward and stuck his teeth in my behind . . . . The pain wasn't the main thing, but the feeling that it was a bit humiliating and unnecessary. . . . Later I myself participated in such a rite, together with the rest of the staff, although no one was very enthusiastic about it. . . . It wasn't a sexual perversion of the captain, but something to do with his will that we become men. . . . It was important to the captain that the officers he's working with are part of the team and prove their masculinity. (Sion 1997: 60; see also *Ha'arretz* daily newspaper, 7.15.94)

This unusual rite contains clear homoerotic elements of the sort that underpin many of the hazing rituals. It is homoeroticism in the form of aggression against "freshmen" (or fresh meat), aggression that is part of the "statusless status" of the liminal period of those trainees who are neither boys nor men.

The National Body in War: Imagining the Chosen Body

In the study I am about to describe, undergraduate and graduate students at the Hebrew University in Jerusalem and Tel Aviv University were asked to visualize their body in two basic situations: the normal body and the body in three states of disease (AIDS, cancer, and heart attack). The directions given to them were, "Close your eyes and try to visualize your body, or your body in disease. Do not interpret or explain, just let yourself be carried in free associations." Most of the students made use of drawing as a medium of free expression.

The project began in 1983 and lasted until 1993. In 1991, it was interrupted by the Gulf War. Both the drawings of the "normal body" made during the eight years before the war (1983–91) and the ones made in the years after (1992–93) present a relatively unchanging pattern; they are representative of the "individualistic" stage Israeli society has entered while preserving many of its collectivist features. The pattern was disrupted in the drawings of the normal body made a few days before the Gulf War and during its course, and then resumed about six months after the war had ended. My argument is that the drawings of the war period presented "collectivist" features—that they articulated more of the public body than of the private. This public body is the internalized representation of a nation in arms, besieged and outnumbered by its enemies. It is a body chosen in order to cultivate and mobilize feelings of solidarity.

Although this chapter is organized around a generalized life-cycle conception, I think this section is relevant here. It is about a specific war, yet apparently wars are always a part of the Israeli life cycle. The body image that is illustrated in this chapter follows Israelis throughout their life. It hovers above the screening of bodies of children, described earlier. But it surfaces most blatantly in war, as the following paragraphs should illustrate.

Let me first say a few words about analyzing drawings in general, especially drawings of the body. The psychological literature on the subject, including the established tests like Draw-a-Person or Kinetic Family Drawing, is of course abundant (see Klepsch & Logie 1982). Drawings are interpreted as projections, in which the drawn figure is in a sense the one who drew it.[14] My analysis will not make use of psychological tests or psychological methods. Instead, I will attempt a more general, nontechnical, and

TABLE 2

Characteristics of Student Drawings of the Body Before,
During, and After the Gulf War, 1983–1993

(Percentages)

Characteristic	1983–1991 (N = ca. 6,000)	Gulf War Jan.–Apr. 1991 (N = ca. 180)	1992–1993 (N = ca. 120)
Body posture			
Standing	90%	79%	90%
Profile	3	–	2
Front	82	79	86
Back	5	–	2
Sitting	7%	10%	3%
Profile	–	4	–
Front	4	6	3
Back	3	–	–
Lying	6%	8%	7%
Profile	–	5	–
Front	6	3	7
Body contours			
Whole	96%	20%	98%
Framed[a]	–	77	–
Fragmented	4	3	2
Face			
Whole	100%	32%	100%
Concealed[b]	–	68	–
Gender marks			
Long/short hair	44%	4%	51%
Lipstick	25	–	22
Clothes	61	10	70
Earrings	11	–	9
Emphasized body proportions	86	4	83

SOURCE: Weiss 1997b.
NOTE: All percentages are rounded up.
[a]Contained by a larger shape, usually a sealed room.
[b]Typically by a gas mask.

sociological approach that considers the figure drawings, in the words of Stefonica Pandolfo (1989: 7), "allusions to something not in the picture, but something else, or at least something implied but not said, . . . fragmented elements of a figurative script that presents obliquely what, perhaps, cannot be represented otherwise."

Table 2 summarizes my findings. One of the most astonishing was the difference between the normal body as drawn in the pre- and postwar periods and the normal body in the wartime drawings. But the latter were strikingly similar among themselves. I therefore termed this new body "the body in war." Many of the students (almost 70%), however, told me that what they drew during the Gulf War was the normal body that I requested, and not something else. The features of the normal body and the body in war are compared in Table 2.

The table can stand some explanation. The main problem was defining the parameters by which to compare drawings. Those parameters had to be "objective," that is, had to represent a graphic feature without reading into it a reified preconception. For example, "passivity/activity" cannot be selected as a primary (independent) parameter, although bodies in war may be "passive" and normal bodies "active," because this is a subjective quality. I therefore searched for more basic, graphic parameters, from which a quality such as "passivity/activity" could be inferred, for example, body posture. Moreover, each graphic parameter is itself "qualitative," that is, divided into several subparameters. For example, posture is minimally divided into three states, standing, sitting, and lying, and each of these is minimally divided into three positions, front, back, and profile. This simple subdivision already generates nine table cells just for posture, each cell with its own distribution per period. In addition, although there can be many significant parameters, I had to select only the most important ones—those that demonstrated the largest intergroup similarities as well as intragroup differences. The set of categories I came up with is therefore offered here as one possible reading of a graphic text in need of verbal decipherment—which is to say, the categories are far from being fixed.

The numbers presented in Table 2 stand for the percentages (always rounded) of drawings that incorporated a specific graphic category during a specific period of years. In regard to quantities, each year, beginning in 1983, I collected drawings from students from three classes, about 20 students per class. This yielded 60 drawings per body type per year, or about 6,000 drawings per body type for the period 1983–91. During the Gulf War, I collected drawings three times (rather than just once a year), because of the special circumstances, or a total of 180.

Looking first at body posture, we see that, as one might expect, both the normal body and the body in war were usually (about 80%) standing and seen from the front. However, only the normal body was shown from the profile and the back while standing. Moreover, where the normal body lying down often suggested a state of relaxation and leisure, its wartime counterpart seemed instead to reflect anticipation.

In the case of the body in war, two postures were especially significant: the sitting and lying in profile. Usually, in these drawings, the figure's gaze is fixed on some communication appliance, such as a TV set, a radio, or a telephone, with the face turned away from the viewer. These positions of passivity are suggestive, reflecting the passivity Israel experienced during the war. The Gulf War did not entail the mobilization of forces; there was no call-up of army reserves, for example. On the contrary, it entailed the demobilization of Israeli society, which was relegated to the home and asked to wait there for instructions (Werman 1993; Shaham & Ra'anan 1991). Radio stations were merged into one channel, and TV stayed on the air 24 hours a day. The home became both a total sanctuary and an information center (A. Ben-David & Lavee 1992). The media appliances surrounding the body in war are symbolic of this passive sit-and-wait state.

In terms of body contours, the normal body was almost always wholly drawn, that is, with full contours. The body in war, in contrast, was contoured more than once: it was framed, contained by a larger shape (77%). This was usually a sealed room. Sealed rooms were prepared in each Israeli home in case of a gas attack (see Werman 1993; and Shaham & Ra'anan 1991). During SCUD attacks, Israelis would rush to their designated sealed rooms as the sirens screamed, frantically trying to complete the insulation procedures and to put their gas masks on. The sealed room, together with the gas mask, therefore became a symbol of the Gulf War. The surroundings in the drawings, however, could also be a house, a city, or skies filled with threatening rockets. In a parallel manner, the face of the normal body was always whole, whereas the face in war was concealed, usually by a gas mask, which was part of the frame discussed earlier. Twenty-four-year-old Miri's description of her drawing of the body in war highlights the body images of that period:

I see a frontal body . . . almost like the normal body, but with a gas mask that covers the face. It has a skirt, so it must be a woman. . . . But the body is kept within a rectangular frame. This is the sealed room. A sealed body in a sealed room. Individual traits of the body give place to the community . . . I think of the community, not of any specific body. Without the army, the politics, the media, we would not have any bodies. All of us are in the same situation, and we are all waiting for the all-clear message from the military spokesman.

The normal body, as could be expected, had gender marks as part of its presentation. Adding hair was very common, and it usually followed the normatively prescribed style—long hair for women, short hair for men. A parallel emphasis on body proportions was also expressed. The masculine normal body featured broad shoulders, the feminine a thin waist, a breast line, and narrower shoulders. Some female bodies included lipstick and necklaces, and masculine bodies might feature chest hair and mustaches or beards. The genitalia, however, were scarcely drawn. Some male religious students added yarmulkes *(kipa)*, and some religious married women a hat. Finally, although some students were as old as thirty-five, the normal body, for them, as for all the students, was always a young body, with no signs of aging. The various marks of gender were almost lacking in the body in war. Clothes were the only exception (10%). However, these were not the normal, gender-marked, variable clothes of the normal body, but almost always training suits, the favorite apparel during the Gulf War (see Danet et al. 1993).

Finally, whereas the normal body was drawn by itself, in focus, and with no surroundings, the body in war was framed within detailed surroundings. These usually included other figures (said to be family members) in the same room, and a TV set, a radio, or a telephone, or all three. In the background, there were sometimes skies with missiles. Sometimes other people, proportionally smaller, were drawn in the background, outside of the framing contours of the room.

The categories of the normal body and the body in war must be considered symbols of a more general social order, embodiments of the dialectic of the collective and the individual. The normal body was presumably drawn during a period of growing individualism, which is a social order characterized by hedonism, free thinking, realism, pluralism, postmodernism, and body care. Some of these characteristics indeed manifested themselves in the drawings of the normal body, mainly through the graphic principles of plural forms, items of body care, the realistic depiction of gender marks, and an attempt to portray one's "self identity."

The framed body in war, its face concealed and often genderless, never alone but surrounded by other persons, and wired to the media, is arguably the concrete image of rising collectivism. The qualities that manifested themselves in the chosen body of the pioneer (halutz) and re-appeared in the body image of the soldier, are captured again during the Gulf War, although in a unique variation reflecting the Israelis' overall state of passivity. The collectivist body image during the Gulf War centers on Israel's military experience; the difference is that it is now an experience of siege. This is manifested mainly through the embodiment of standardization, passivity, sameness of form, faceless features, the body's dependence on external frames, and its state of ordeal.

War, it should be emphasized, subjugated the individual body by masking its face and effacing its individual features. This subjugation resulted in a focus on what could be termed the external body. To make this point clear, let me say a brief word about the masking of the body in war compared with its masking in a completely different situation—disease. As mentioned, I also asked my students to draw the body in three states of disease. In the AIDS and cancer drawings, in particular, the face, as in the body in war, tended to be masked, but now with a skeleton face. Body contours were also masked, but not with external frames. AIDS and cancer had a body whose skin no longer covered it, no longer provided protection. It was a body without an immune system, in Martin's terms. Bodies in war, therefore, were framed by larger forces, maximizing their external boundaries as sources of protection—and losing their self-contours as a result of this submerging in the surrounding community. Bodies in AIDS and cancer, in contrast, lost their contours because of an inner contamination. Whereas the reference point in the case of war was uniformity and totality (hence, collectivism), the reference point in this case was nullification.

Interestingly, the "collectivist" pattern of drawing the body, which appeared just before the beginning of the Gulf War, endured well after its termination. A month after the war, though students reported to me that they now had now forgotten their fears, and that "Sadam's biochemical missiles did not exist," they still depicted the body in war. It thus seems that the mode of collectivism, invoked by states of war, is too powerful a conditioning to easily evaporate immediately after the threat is gone.

The Chosen Body and the Institute of Forensic Medicine

Located in Tel Aviv, the Institute of Forensic Medicine is an intriguing meeting ground for different, almost opposite approaches to the body. On the one hand, it is a scientific institute, affiliated with the Sackler School of

Medicine (Tel Aviv University) and operates a state-of-the-art genetic laboratory. On the other hand, it is closely inspected (although informally) by the Chevra Kadisha (Aramic for Holy Society), the religious organization that, except for the army and kibbutzim, has a monopoly on burials. On the one hand, the institute is a civil organization working under the Ministry of Health. On the other hand, it fulfills the requirements of the military and the police. I conducted observations there during 1995–2000; the following summarizes my findings in regard to the chosen body.

The Institute of Forensic Medicine conducts tests in order to identify bodies and the results of physical violence. As the only such institute in Israel, it receives thousands of cases a year. These cases include rape, medical malpractice, infant deaths, the battering of jail inmates, and the deaths of Palestinians in security interrogations (Hiss & Kahana 1996).These cases are brought to the institute by various state organizations in need of a professional opinion, such as the police, the IDF (Israel Defense Forces), the prison authority, and the Ministry of Health, as well as private parties. Autopsies are performed only with a court order or with the consent of relatives.[15] Despite having been created to support the state authorities, the institute operates as a guardian of human rights, often in the context of male and female minorities (prisoners or terrorists, ignored rape victims, and others).

From a sociological point of view, the institute provides a unique meeting ground for the personal body and the social order. The cases it examines, the nature of its practice, and the applicability of its findings are directly connected to breaches in Israeli society, in particular to nationalistic, religious, and ethnic breaches. The institute maintains (often unwillingly) different standards for Jews and non-Jews, Palestinians and Israelis, soldiers and civilians. These distinctions have received a lot of sociological attention, but never in specific reference to forensic medicine.

The link between forensic medicine and social rifts has been documented from various perspectives and in different geopolitical contexts (see Tedeschi 1984; Asad 1997; and Jadresic 1980). Amnesty International and the American National Academy of Sciences, for example, have published a lot of material on human rights violations in Guatemala (Amnesty 1989, 1990; NAS 1992). Recent years have brought huge advances in forensic medicine, such as DNA and tissue matching, both of which have for some time been considered valid by the courts of most of the developed countries. In Israel (as elsewhere), it is these sophisticated technologies, as much as anything else, that have cast Israeli physicians into their new role as guardians of human rights (Gordon & Marton 1995).

But this role has created some problems in Israel, because physicians are not mere neutral agents of biomedical science. Together with the Israeli family, Israeli medicine is a major agent of socialization. Physicians are the guardians of Israeli health and selectors of the chosen body. They perform that job as obstetricians and gynecologists (as described earlier), as gate-keepers for elite military units, and as forensic physicians. Several of the physicians I interviewed mentioned these conflicts of interest.

Forensic medicine has a special role in policing. Historically, it has been largely constituted as a judicial aid. The "Outline of Forensic Medicine," written by the physician Johann Valentin Muller in 1796, "exemplified the attitudes of medicine toward those who sinned against convention" (Mosse 1996: 29). For example, forensic physicians were occupied with identifying outward signs of homosexuality and excess masturbation (onanism) con-demned and prohibited by the Church and the state alike.[16] According to Mosse (1996: 27), forensic medicine came to the aid of judges and juries try-ing to enforce the laws against sodomy by developing a stereotype for use in identifying homosexuals. Homosexuality was viewed as dangerous to the health of the state. Ambroise Tardieu, in his *Crimes Against Morals from the Viewpoint of Forensic Medicine* (1857), stressed a "feminine appearance and diseased body" as outward signs in a male homosexual. In Israel today, physicians serving in the army medical corps similarly serve as gatekeepers of masculinity by selecting and grading prospective recruits and further selecting those who volunteer for the elite combat units.

One of my most startling findings was that the handling of bodies in the institute reflected the boundaries of collective identity in Israel. Jews versus non-Jews is the first dichotomy when a potential chosen body, which is to say, a man's body, is examined. If it is not circumcised, all procedures are halted until the staff can determine whether or not the dead man is regis-tered as Jewish. Mistakes are known to occur, and the staff takes special care not to mislead Chevra Kadisha so that non-Jews get buried in a Jewish cemetery. If it is found that the dead man is Jewish, but for some reason is not circumcised, members of the Holy Society often take it on themselves to circumcise the body. There are additional ways to "correct" the Jewish body after death. For example, tattoos related to Christianity (usually a cross) are removed. The procedure is performed by Holy Society workers, not the institute's staff. Another prominent example of correcting the Jewish body, or keeping it perfect, is the returning of tissues to the grave. According to the halacha, "the blood is the soul," and therefore blood dripping from the body is collected into pieces of cloth, which are taken to the cemetery to be buried with the corpse. Every week, an elderly employee of Chevra Kadisha comes to the institute. His sole responsibility is to collect tissues that were taken from the body for histological tests and return them to the grave. These three procedures—circumcision, tattoo removal, and tissue burial—are performed only for Jews.[17]

An even more stringent dichotomy obtains between soldiers and civilians. The bodies of soldiers are kept apart and handled almost ceremonially. It is forbidden, under all circumstances, to take body tissues from soldiers. Many of the institute workers stressed in interviews how soldiers should be kept "integral" (*lelo dofi*): "the soldier is a hero, his body is sacred. We must not touch it, we must not take away anything," said one. "His mother must be able to recognize him," another insisted. "He is one of us." Broadly speaking, the ceremonial handling of the soldier's body follows the traditional Jewish commandment to honor the dead (*kvod ha-met*). In this manner, there is a continuity between Jewish and Israeli customs as far as the dead body is concerned. More precisely, Israeli culture (as perhaps most clearly exemplified in the institute) follows and even adds to the Jewish tradition as far as soldiers are concerned. One of the assistant pathologists told me that for soldiers he uses special skin-color thread to camouflage any stitches.

The desire to keep the fallen body as complete and intact as possible is reflected in two prohibitions. It is forbidden to harvest skin from soldiers' bodies, even when the available skin meets the required standard, and even though this operation might save the life of other soldiers. The national skin bank at Hadassah hospital, which is designed to cater to military needs, does not receive any of its material from soldiers' bodies.[18] Soldiers' bodies also cannot be used for the development of medical skills. Prof. Arie Eldad, head of Hadassah's plastic surgery department and former chief of medicine in the IDF, told me that the institute had long used bodies to practice on, and he had only recently learned that it never used soldiers' bodies for that purpose. This was not ordered from above, but rather reflected "the institute's own policy." Professor Hiss, the head of the institute, told me that he had not issued any such guideline, because it "would have created a hierarchy of bodies." Although it is uncertain where the policy originated, in practice soldiers' bodies are not practiced on, despite the fact that the skills learned might save the lives of other soldiers.

In 1995–97, a new social issue rose to the top of the institute's agenda. In what the media dubbed the case of the "missing Yemenite children," reports surfaced that there had been a wholesale kidnapping of the children of Yemeni immigrants during the early 1950's for the purpose of putting them up—without their parents' knowledge or consent—for adoption by Ashkenazi families. Many immigrants from Arab countries were flown to Israel before and after 1948. Among these were families from Yemen, who were gathered in transition camps. Following many complaints regarding the disappearance of Yemenite children from hospitals and schools within these camps (in 1997, the missing were estimated to number around 1,000), a state commission of inquiry was established in 1995 to look into the affair. It authorized the Institute of Forensic Medicine to open the graves of adopted (Yemeni) children and conduct DNA tests in order to establish the corpses' "real" lineage, that is, to match them to their alleged Yemenite families. The process of exhumation and identification began by taking blood samples from ten chosen families and comparing them with the DNA of the corpses.[19]

I find the affair of the missing Yemenite children interesting as another example of the Israeli discourse on the chosen body, and particularly the move from one collective, interchangeable body to many different, ethnic bodies. The affair was constructed as a burning ethnic issue that threatened to dismantle the integrity of the body politic of the state and its (Ashkenazi) elites. The protest of Yemenite activists, and the forming of a state commission of inquiry in response to that protest, imply that what could happen in the early 1950's—at the height of the melting pot doctrine, of high collectivism, of Ben Gurion's statism—is delegitimized today. The story of the coerced adoption of the Yemenite children is revolting evidence of the melting pot doctrine, of the intent to assimilate all Jews of different origins by effacing ethnic traditions and subjecting them to Ashkenazi hegemony.[20] It is both ironic and fitting that the exhumation of ethnic identity should be performed by the institute, that great gatekeeper of Jewish identity.

Another notorious case is worth mentioning in this connection. On

January 25, 1996, a report in the Hebrew daily *Ma'ariv* revealed that for years the Israeli blood bank had been quietly destroying blood donations made by Ethiopian Israelis. The reason given by officials was the high rates of HIV, hepatitis-B, and malaria within that community, from which the general public needed protection. The Ethiopian community, which suffers a marginal social standing, demonstrated against what it regarded as another form of exclusion.[21] "One People, One Blood," was among the most popular of the placards the Ethiopian demonstrators carried (see Seeman 1997: 197–205 for a full description of the incident). Demonstrators complained that for the Israelis, Ethiopians were good enough to be killed as soldiers, but their blood was apparently not as good as Israeli blood. Another popular slogan chanted at the demonstration was "We Will Not Let Our Blood Go Ownerless" (*lo niten damenu hefker*), which meant, as Seeman interpreted it, that Ethiopians would no longer tolerate being less than full Israeli citizens. One of the demonstrators, a recently discharged soldier, told Seeman (p. 202) that he had donated blood "every day" while in service. (The IDF is indeed a common source of blood; 33 percent of all donations are collected at army bases; Navon et al. 1996: 11.) "We are as Jewish as the Yemenites," another demonstrator with a loudspeaker shouted, "and more Jewish than the Russians!" Gathering steam, he went on: "They all think of us as sweet Ethiopians. Today we have come to show them a different face. . . . If it takes violence, then we will use violence. We will raise them up another Uzi Meshulam!" (Uzi Meshulam was the well-known leader of the Yemenite Jewish activists who protested against the kidnapping of the Yemenite children. To summon up his name was thus to lend power to the Ethiopian struggle.) The "blood affair" of the Ethiopians illuminates in a concrete manner how the national body is defined in Israel through personal bodily constituents—here, Israeli blood.

My sister Ofra died as a young girl from polio in the 50's. Today (1999), during a conversation with Ora Shifris (the spokesperson of Uzi Meshulam, the Yemeni leader), I tell her about my sister, how she got sick, was hospitalized, and died the same day. I was always told how she was "a healthy child in the morning and dead in the evening." "There are similar stories about the dead/missing Yemenite children," she said. Ora told me that there can be no such thing as hospitalization and death from polio in the same day. Polio takes at least ten days, she said. In the 50's, according to Ora, "the hospitals were crooked." "Don't believe them; the doctors took your sister like they took the Yemenite children. She is probably alive. Give me all the details, her identification number, and I will ask our computer experts to look her up for you. You have a sister."

I didn't know what to do about this conversation. I asked my father if he saw Ofra's body. "No," he said, "but I have one memory of her in the last days: I came to visit her, in the hospital, and they didn't let me in. I had to push my way into the room. I saw her, alone, sad, and when she saw me she began to smile. Then they took me out. Brutally. When she saw I was leaving the room, she looked at me with a face that told me she knew she was losing me forever. I will not forget that look." He told me about this 50 years after it happened. Not a minute sooner.

At this point, I think about the parents of the Yemenite children; how their children were separated from them, how their children became sick and were taken to the hospital, and remained there until the death notice. I think about the brutality of a medical establishment, that didn't allow parental visits because of some "risk of contamination" or "lack of order." I realize the strength of that establishment, the medical establishment, in relation to individuals, even hard-headed Ashkenazi individuals like my father. I think how tempting it must be to believe that your kith and kin are alive. I remember how I used to daydream for hours that my sister would appear again. I used to play with paper dolls, dress them up, and hope they'd come to life and be Ofra. I am almost tempted to give Ofra's details to Ora Shifris.

[From my diary]: Tami, my daughter, is undergoing an officer's course in the IDF. I am ambivalent: on the one hand, worried sick, recalling my own experience of ordeal in that course; on the other hand, happy for her, and in a strange way very proud and anxious to see her through. I wait for her at home on the weekend, wondering what she'd look like. Hoping that she'll come in the door, radiant, and say how good she's feeling. Worried that I'll act like my parents during my officer's course. I'm afraid of her weakness and of my impotence. You raise a child for 18 years, and one day they take her away from you. Just like that. And all sorts of nineteen-year-old children decide what will happen to her. If she's fit or not. When she gets sick, you cannot take care of her. You cannot even take her to a doctor. Nobody wants to hear from you. She's a soldier. What do you mean she's a soldier—she's my daughter, isn't she? I feel bad. I feel bad for Tami. I don't want her to suffer in the army like I did. I'd like to fall asleep for the entire summer and wake up when the course is over.

[From my diary, a letter to Tami that was not sent to her]: "My child. I love you so much. I don't want you to suffer. I know how difficult it is. I know what I've been through. Why do I push you into such a place? You call home, and you cry to me. And I fall apart. In the morning I have to go to the publishers; my book about parenthood has just been published. I need all my strength to go out the door. In class I teach about parenthood. I can barely speak. I need to hold you. When was the last time I held you?"

The letters I actually send her are different: cheerful, happy. The following one, a typical letter, was written and sent after a visit in the camp where Tami was. During the visit, I felt sick and nauseous. My body remembered my harsh experience in the IDF. But the letter was happy: "Hi, Honey, . . . It's really difficult, but I'm sure you can handle it. I saw you yesterday, and it made me calm down. I understood what a great daughter I have. I have complete faith in you. But with all due respect, an officer or not, even if you won't finish this course, I know what you're worth. Not making it is not the end of the world."

Reading back through these letters, I feel what Simone de Beauvoir called a sense of dédoublement, of being double. My body remembers the pain, while my mind is rationalizing. I am nauseous, but take the ride again, despite all the alarms. I am mother no. X, just like one of my respondents in this chapter, taking part in the cult of the chosen body, speaking in two languages, the passive and the active. The only difference, I think to myself, is that I write, too. But does writing make a difference?

3 SANCTIFYING THE CHOSEN BODY

BEREAVEMENT AND COMMEMORATION

IN FEBRUARY 1997, a tragic helicopter crash killed 73 Israeli troops on their way to south Lebanon. A national day of bereavement was declared by the government. Politicians, dignitaries, and the media all re-enacted their sad routine of mourning, consoling, publicizing obituaries. An investigative committee was quickly assembled and started its work. A few days later, the well-known Israeli sociologist Baruch Kimmerling published an interesting criticism in one of Israel's mainstream daily newspapers. Entitled "A Moment of Solidarity," it reflected on the collective discourse of bereavement as a national reflex:[1]

> It seems that the more divided our society is, the more we need such moments. . . . The generation who did not experience the magical days of anticipation just prior to the 1967 War, or the lingering national depression of 1973, could now renew (for a little while) the holy days of the "candle children" who appeared—albeit for a short time— after Rabin's assassination. . . . Not that one should doubt the sincerity of people's feelings. The trauma and mourning are real, even when disseminated by anchor persons. . . . In such moments, the voice of reason must remain silent. . . . When such moments [as this] are to a large extent prescribed from above, by the elites, the public—or parts of it—easily lends itself to be manipulated. (Kimmerling 1997: B2)

This chapter examines and extends the point Kimmerling is making. It does so by looking at many commemorative moments of solidarity, not only those emanating from unexpected, tragic events, but also those prescribed in the national calendar and institutionalized in the social fabric of the nation. Bereavement and commemoration, in Israel and elsewhere, are powerful social discourses. They constitute a symbolic space where the public and the private, the military and the civil, inevitably meet. This space is therefore traditionally marked by normative control, from the smallest burial ceremony to the largest war memorial. Control is particularly evident in the detailed prescription of how to bury, grieve, and commemorate "the fallen soldiers." Here, the personal loss of individual bodies is directly linked to the central myths of national strength and regeneration.

My objective in this chapter is to explore how bereavement and commemoration are appropriated by the Israeli collectivity and reproduced in public life so as to sustain collective boundaries and national ideology. Commemoration exemplifies how the collectivity generates its embodiments—utopian bodies with no gender or individual faces: a body in which Ashkenazi and Sephardi, male and female, those killed in battle and those killed in training accidents, should all become one (as in the case of the pioneer or the sabra). However, certain Israelis—notably Israeli Arabs, non-Jewish Israelis, and ultra-religious Jews—are excluded from this commemorated body. My argument, in a nutshell, is that commemoration is a collectivist-nationalistic project privileging a chosen Jewish body, linking a Jewish past with an Israeli present. In addition, it is quite opportunistically used for social mobility.

Commemoration and Bereavement:
An Anthropological Overview

The social uses of funerals and mourning have been traditionally studied by anthropologists as universal group mechanisms for the enhancement of solidarity.[2] This perspective has been extended to the context of fallen soldiers, whose bereavement and commemoration often take the form of a national "hero-system" (Becker 1971), where the "symbolic immortality" (Lifton 1977, 1979) of the fallen reaffirms the sanctity of the homeland and the hegemony of the collectivity (Mosse 1990; Whaley 1988).[3]

Today, the fallen soldier is perhaps nowhere more publicly honored than in Israel. From the very start, with the thousands who paid with their lives in the War of Independence in 1948–49, Israeli society has deemed such personal losses as essential to the survival of the Jewish state in the promised land (Palgi & Durban 1995). The life sacrifice of young men has been a central, recurrent, and usually taken-for-granted issue in Israeli literature, endowing "the fallen" with an aura of symbolic immortality (see Nave 1993; and Miron 1992). All this has provided a fertile ground for social inquiry, which has recently grown more critical in nature.[4]

This chapter does not purport to exhaust the discourse of commemoration in Israeli society. The discussion is confined to three significant fields. I concentrate first on two important state institutions, the Ministry of Education and the Ministry of Defense and its affiliated Rehabilitation Department, as well as the national association of bereaved families, Yad Labanim (Memorial to the Sons). Yad Labanim and the Rehabilitation Department are closely, almost symbiotically, connected (see Katz & Eisenstadt 1973; and Shanigar & Handleman 1986: 25–27) These organizations issue commemorative materials and are responsible for carrying out burial ceremonies and rites of commemoration and providing the continuing financial support to bereaved families designated by law. The chapter next turns to bereaved families and individuals, some of whom have recently organized to challenge the ways in which the establishment treated the deaths of specific soldiers. The final topic is less tangible but no less significant: it is commemoration as disseminated through the media, in television, radio, and newspapers.

"The Silver Platter": Commemoration as a Scholarly Program

Memorial Day for the Fallen (Yom Hazikaron) is the national day of remembrance for soldiers who have sacrificed their lives for Israel. Its ceremonies are performed at schools throughout the country seven days after Holocaust Memorial Day (Yom Hasho'a), and they conclude exactly one minute before Israel's Independence Day (Yom Ha'atzmaut). Memorial Day for the Fallen (MDF, for short) is therefore part of a larger narrative—one leading from destruction (holocaust) via sacrifice (the fallen) to salvation (independence). Instructions for conducting the MDF school ceremony have been issued each and every year since 1954 by way of a general memorandum from the Ministry of Education. This annual reissuing of what amounts to virtually the same instructions is one example of the cyclical time underpinning the mythical narrative of commemoration. Memorial Day for the Fallen is about symbolic immortality.

The state ceremonies enacted on these three interlinked days are held in the same general manner in every school in the country and are attended by the highest officials and dignitaries.[5] The schedule, order, and contents of the ceremony are all prescribed by the Ministry of Education. It should be mentioned that the ministry does not supervise the ultra-religious schools.[6] School ceremonies are plainly a primary educational means for inculcating the younger generations of Israeli citizens with the messages of state ethos and national identity (Shapira & Hayman 1991; Goldstein 1980). MDF therefore figures prominently as a "scholarly program."

The ministry's MDF instructions are rather laconic. They generally order that

> on the fourth of Iyar,[7] Memorial Day for the Fallen, at 7:00 A.M., the flag of the state will be flown at half-mast. It will be raised again at sundown. On this day, a solemn ceremony will be conducted and attended by all teachers and pupils, starting at 8:00 A.M. with a siren proclaiming a two-minute silence during which all activity and traffic cease. The MDF ceremony should be conducted so as to bridge between sorrow and joy, grief and celebration. . . . It is a day on which the living rededicate themselves to the State of Israel, so that they may be worthy of the sacrifice of those who died for its survival. It is desired that former graduates and pupils' parents, as well as bereaved parents whose sons were pupils at the school, be encouraged to attend the school ceremonies. (Ministry of Education, General Memorandum, 31.3.1957, p. 3)

The instructions, though usually merely repeated, are not invariant. For example, for the MDF after the 1967 war, the ministry referred to the recent events by stating that "MDF's meaning will be clearer to the pupils this year. We mourn today the death of our heroes, and feel proud of their heroism and devotion. The champions of Israel went to the battle of freedom with open eyes, and in their death they commanded us life" (General Memorandum, 30.4.1968, p. 10). The last sentence is a powerful ideological expression, one that has become a motto in the national reference to the fallen. It reappears in various semantic and symbolic forms in the texts read during the ceremony. The systematic construction and reproduction of the ideological nexus between death and life, sorrow and joy, remembrance and independence, is one of MDF's most significant feats. MDF is therefore of utmost importance in the reproduction of an "ethos of sacrifice." With this term, I mean to say that the death of a soldier, irrespective of when, where, and why it occurred, is construed as a sacrifice (i.e., justified as serving the national cause). It is an ethos in the original Greek sense of a tragic deed that is ethically justified because it was predestined by a larger moral cause. Death is constructed as necessary for national life. In such a way, the fallen are at once endowed with symbolic immortality and depersonalized. Their death is cleansed of its individual features and possibly messy circumstances. It is therefore only natural that celebrations of Independence Day immediately follow the MDF. On the same day, 6:29 P.M. is still considered remembrance, which turns, at 6:30, into independence.

MDF's concrete prescriptions for commemoration services further deepen the depersonalization of bereavement. MDF is also a means for defining collective boundaries. Its target population is the whole nation, not just the bereaved families. Effacing the differences between those who have lost a son and those who have not, it transforms the whole nation into "one bereaved family" (a favorite phrase of politicians).

Following is a concise description of a typical MDF ceremony, as I observed it at Chestnut Hill School in Tel Aviv in 1997. Behind the stage, erected especially for the ceremony, there was a black backdrop emblazoned, in white, "Remember" (*yizkor*). The stage was flanked by a choir and a "remembrance torch," which was lit during the ceremony. At center stage, a teacher and ten pupils stood grouped around a flagpole flying (at half mast) the national flag, all of them (like the rest of the students) wearing white blouses. At 8:05, the ceremony was called to order by the teacher. A pupil then recited "*Hatzvi Yisrael*," the opening verses of David's lamentation to Jonathan. Following this, the principal gave a short speech. A pupil read the secular memorial prayer, *Yizkor*; the choir sang a song addressing a younger brother who fell; a young girl stepped forward from the choir to recite a short prose piece regarding the significance of MDF; an older child recited a piece written by a soldier who fell; another song was sung; and so on. That all the songs were well-known, classic Israeli creations was evident in the audience participation, many of the crowd humming along with the performers. The ceremony culminated in the reciting, by the whole choir, of "The Silver Platter" ("*Magash Hakesef*"), an Israeli poem by Natan Alterman centering on the sacrifice of young soldiers, whose lives were the "silver platter" upon which the state was served. Here are the first and last verses:[8]

> Unwashed, weary unto death, not knowing rest,
> But wearing youth like dewdrops in their hair
> —Silently the two approach
> And stand.
> Are they of the quick or of the dead?
>
> . . .
>
> Through wondering tears, the people stare.
> "Who are you, the silent two?"
> And they reply: "We are the silver platter
> Upon which the Jewish State was served to you."

The two soldiers—male and female—share the collective features of the sabra archetype. They are young, hard-working ("in work garb and heavy-shod"), determined, natural ("unwashed and weary"), silent yet straight-talking. They bear no individual marks, no names or gender traits. Death did not change them, only made them more sabra-like. The bodily features idealized by canonical MDF texts are the features of the ideal, collective sabra. The fallen and the living sabras all have "the same physique, the same high-rise hair" (in the words of Naomi Shemer, from "We Are from the Same Village"). The fallen, like ideal sabras, come "in a masculine step, strong and sun-burnt" (in the words of Hayim Hefer, from "The Parade of the Fallen").

In the Introduction, I emphasized that the cultural image of the sabra was primarily defined through the body. The palmach fighter, who became the authentic representative of the native-born, is often described as natural, simply dressed, tanned, tall, and strong, and with "high-rise" hair. The sabras are wild, untamed, and unstyled, yet pure and sincere. The fallen reflect and strengthen that image. They are pure sabra. They must be, since they are the national sacrifice. According to ancient Jewish tradition (which is probably common to many human societies), the sacrificial offering to God—be it an animal, a food, or a valuable—must be the best and purest of its kind. This precondition underpins the meticulous regulations of the halacha, as well as the biblical melodrama of Isaac's sacrifice (*akedat Yitzchak*). Uri Zvi Grinberg, a poet who stressed Jewish nationalism, wrote that the fallen are "The Chosen . . . the true sons of the race of David" (from "Those Who Live by Their Virtue Will Say").

The sacrifice myth that connects the Israeli fallen soldier and the biblical story of Isaac was introduced in the very first of the Hebrew commemorative albums bearing the title *Yizkor*. According to Oz Almog (1997: 74), the Israelis made three changes in the original biblical myth. First, Isaac's passivity—being sacrificed by his father—was turned into activity, as the soldier voluntarily sacrifices himself. The sacrifice thus becomes an active story of heroism rather than a passive tale of abiding. This active heroism was further emphasized by the fact that the new martyrs die with their weapons in hand, in battle, defending their country. "Theirs is bodily heroism rather than *kidush hashem* (dying for the lord)" (ibid.). Second, the sacrifice was now made to be instrumental for the fulfillment of the Zionist ideal and a test of tribal loyalty, rather than to please some vague deity. Third, and most important, the focus was changed from the relationship between Abraham and his God to the relationship between Abraham and Isaac. The latter relationship is the one between the generation of parents, the pioneers, and the generation of sons, the sabras. The commemorative literature is filled with generational references to the fathers and the sons. "Blessed be the parents who had the privilege of raising such sons! Blessed be the people whose sons sacrifice themselves on its altar!" asserts one of the early commemorative books (cited ibid.). The palmach prayer *Yizkor* (for remembering the fallen) proclaims: "We extend a courageous hand to the fathers and mothers who sacrificed their sons on the altar of the homeland" (ibid., p. 75). Chaim Guri, in one of his famous bereavement poems, writes:

> Isaac, we are told, was not sacrificed.
> He lived many days,
> Saw the good, until his eyes became dark.
> But that hour he left for his descendants.
> They are born with a knife in their heart.

"That hour," the hour of sacrifice, is the sabras' inheritance. The sacrifice—as inculcated by the Zionist ethos—has become the genetic code of the Israeli soldier, who is "born with a knife in his heart." The sabra hence represents a generation of sacrificial "lambs" or the "silver platter" on which the state of Israel was constructed. The sacrifice, a traditional Jewish concept, thus became the genetic code of the new Israeli culture of commemoration. The Zionist son abandons his father in the diaspora, where the father is going to burn in flames (the holocaust); and the pioneer who became a father brings his son to the altar.

In the poetry of the 1940's and 1950's, for example in Chaim Guri's poems, the sacrifice is a lingering condition in the nation's life. The return to Zion entailed a return to the necessity of the sacrifice, this time as part of nation-building. As the poet Yitzchak Lamdan wrote, "And here [in Israel] we were all sacrificed and with our own hands we brought the wood" (from "On the Altar"). Sacrificial images can be found in many other poems written during and after the founding of Israel. For example, in the poem entitled "Voices" by Tuvia Rivner ("I know, my son, I am the father. / I am taking you with me, we both walk together"), in "Fear of Isaac" by T. Carmi ("Tonight I dreamt that my son did not return"), in "Sara's Life" by Binyamin Galai ("And the coffin in which she laid was made all these years / of memories of broken wood on another mountain / another mountain, in the land of Moriya"), in "No Deer to Replace Him" by Ayin Hillel ("But your son, your only son, whom you love, / There is no deer to replace him"), in "Sacrifices" by Shne'or, written in the midst of the 1948 war ("And no angel was there to stop the slaying / and no deer caught with its horns in the bush"), and in "The True Hero of the Sacrifice" by Yehuda Amichay ("The true hero of the sacrifice was the deer"; see Karton-Blum 1996). The images of the sacrifice are obviously varied, but even this variety reflects the utmost significance given to that concept in the period of nation-building.

The *Yizkor* albums issued in memory of the fallen represent the direct linkage between sacrifice and the military ethos. The first commemoration book (1911), which has come to spawn a whole genre, proclaimed that "unless we watered the land with our blood, we would not be standing on it today. . . . Without the stamp of blood no national hope has ever reached its goal. . . . The land of Canaan, after a break of hundreds of years, has once again set out to taste the hot blood of her children" (cited in Elboym-Dror 1996).

A perfect body is but one feature of the chosen-ness of the sacrificial offering. This bodily motif is clearly connected to ancient (biblical) Jewish tradition and folklore. Mary Douglas, who has speculated about why the body so preoccupied the ancient Jews, claims that "the Israelites were always in their history a hard-pressed minority. In their beliefs all the bodily issues were polluting, blood, pus, excreta, semen, etc. The threatened boundaries of their body politics would be well mirrored in their care for the integrity, unity and purity of the physical body" (Douglas 1966: 124, cited in Eilberg-Schwartz 1992: 22). The canonical texts of the MDF perform a similar function in disguising the "pollution" of death. Death is bloody and messy, and it usually ruptures the body. All this is missing from the depictions of the fallen in the MDF canonical texts.

The MDF ceremony, then, exemplifies a collectively realized performance within which national ideology can be maintained and reproduced. In this, it resembles many other national school ceremonies, such as the morning recitation of the Pledge of Allegiance in American schools (Middleton & Edwards 1990: 8). In anthropological terms, these ceremonies are ritualized practices for consolidating group solidarity through contact with the "spirits of the fathers" (V. Turner 1977; Moore & Myerhoff 1977). As Ilana Bet-El and Avher Ben Amos (1992) contend, MDF is a theatrical performance in which the performers directly address the audience, without any attempt to create a realistic illusion. Performers and audience become one in a binding myth of sacrifice and martyrology, not unlike the agitprop plays staged in the Soviet Union following the 1917 revolution and in Germany by left-wing groups during the 1920's. The resemblance is clearly no accident, for this theatrical tradition was in fact brought to Israel by immigrants from the USSR and Germany (Offrat 1990).

The myth of the sacrifice (*akeda*) has been described here from one viewpoint, that of the national, collectivist script for commemoration. Yet the akeda has had multiple readings, some of them subversive, within Israeli society, particularly among political satirists. Such interpretations can be found, for example, in Amos Keinan's *Friends Talk About Jesus* and in Hanoch Levin's satirical plays (see Alexander 1985: 142–53). In the short poem "Dear Father," part of the famous play *The Bath Queen*,[9] Levin has the speaker, a fallen soldier, recite the following lines:

> Dear father, when you stand on my grave
> Old and tired and very lonely,

And you'll see how they bury me in the ground
While you stand atop of me.
Don't you stand there so proud
And don't you raise your head, Dad.
Here we are now, flesh against flesh
And this is crying time, Dad.

So let your eyes cry over my eyes
And do not keep silent for my pride.
Something that is more important than pride
Is laying now at your feet.

And don't you say that you've made a sacrifice,
Because I was the one who made it,
And don't you say high words
'Cause I'm already very low, Dad.

Dear father, when you stand on my grave,
Old and tired and very lonely,
And you see how they bury me in the ground –
Ask me then to forgive you, my father.

   This critique, for all its acerbity, serves to demonstrate the dominance of
the collectivist script, represented by the image of the father. The poem is
part of a satirical review that deals with the sacrifice of Isaac. The play provoked a furious public debate and even violent protests. It was an example
of the intolerance of the Israeli public in the 1970's toward such "betrayals"
of the sacred ethos of commemoration. Some 15 years later, Levin once
again made himself the target of debate with his play *The Patriot*, dealing
with the Lebanon war. The following lines are from one of the sketches in
that play: "the classical recipe for independence: / take some people and sentence them to death." I return to the subversive voice of Israeli artists in
regard to bereavement later in this chapter. I now turn from the national,
official script for bereavement and commemoration to the semiofficial narratives of bereaved parents.

Yad Labanim: The Authorized Version of
Personal Bereavement

Yad Labanim (Memorial to the Sons), the national association of bereaved families, serves as the primary representative and lobbying agent for those who have lost their sons in war. As a national association, its "front-stage" texts (such as its bulletins) present the desired, conformist mode of bereavement. That mode is collectivist in nature, what I termed in the above subtitle the authorized version of personal bereavement.

Yad Labanim (YL) was established in the mid-1960's and began to issue annual bulletins to its members at the beginning of the 1970's. The scope of this chapter does not allow an exhaustive analysis of the many bulletins that have accumulated over the years. I have therefore chosen to begin my presentation with an account of a relatively recent debate among YL members. In 1989, the Ministry of Defense published a memorial album containing photos and descriptions of various war memorials in Israel, for dissemination to bereaved families on the eve of the MDF. Tom Segev, a journalist and historian, reviewed the album in the daily newspaper *Ha'arretz* (19.5.89). Calling it the "Catalog of Macabre Israeli Tombstones," Segev criticized the album for reproducing a nationalistic fetishism of a piece with that displayed in the chauvinistic European war memorials that George Mosse (1990) describes. The attack provoked a host of letters to the editor from YL members, which formed the main body of its July 1989 bulletin. Segev's challenge was met with an equally determined response, whose vigor discloses the basic ideology of Yad Labanim.

Although the various replies were different in style, length, and emphasis, they all made use of several recurrent arguments. These could be summarized in the following manner. First, the uniqueness of the bereavement experience. Many writers began their letters with something like, "I do not know Mr. Segev personally, but it seems to me that he cannot see the album through a bereaved person's eyes." In other words, since Segev had not lost a son in battle, he was incapable of reviewing anything that was for and about the bereaved. "War memorials cannot be a subject of art criticism," one writer argued. Another claimed that "Mr. Segev sees in every memorial a cold rock. I see a flag in every memorial. I breathe heroism from these memorials. It enables me to keep my back straight" (1984 bulletin, p. 6)

This rhetoric of uniqueness is typical of much of the bulletins' material. Many pieces could have attested to this fact, but the following poem, entitled "To the 'Others' " and written by a bereaved mother, is perhaps the most forceful: "I am no longer like you / I am something from a different planet / Bereavement flowing through my body / Mourning the primary constituent of my blood / And so forever" (1985 bulletin, p. 4). Intriguingly, the poet's claim for uniqueness is grounded in her body. Her body, the most personal and private of entities, is also the locus of "her" bereavement. By using a rhetoric of uniqueness, so reminiscent (in Israeli eyes) of the holocaust survivors' rhetoric of coming from a "different planet," YL members could be seen to appropriate and legitimize an autonomous discourse of bereavement, one that intends to demarcate their group from "the others." The fact is, however, that YL members are often the main carriers of the state's collective ideology. The private body—as an icon of the intimate relationship with bereavement—is used in specific cases where group solidarity is enhanced through a dispute with "the others" (Tom Segev, in this case). More often, the private body makes way for a generalized body, a body politic, consisting of the entire "people of Israel." The rhetoric of solidarity capitalizes on the ethos of sacrifice in order to turn bereavement into a source of national obligation and personal pride.

The Segev case exemplifies both rhetorical strands. Additional arguments used by YL members in reply to Segev were all clear-cut expressions of the collective ideology of bereavement. Their letters emphasized the "glory, pride, and honor" evoked by war memorials, which "endow the loss of our children with meaning." This is a retelling of the national ethos of sacrifice. Memorials are a "sacred place," which Segev chose to desecrate. Many letters noted the necessary role of memorials in forming a locus of identification between the bereaved and the army—places where both the bereaved and soldiers in active service come to during MDF. For many families, the memorials therefore denoted a feeling of oneness, of "us," a sense of solidarity bought in blood.

YL members have traditionally sided with militaristic Israel, the "nation in arms" (or better, "nation in bereavement"), rather than taking advantage of their social license to promote leftist calls for peace, for example. This is not to say that the organization itself has expressed an explicit hawkish political stance. It has always attempted to stay out of politics by clinging to a depoliticized consensus. When members attacked "the others," it was always in the name of the collectivity. It is interesting to find in the bulletins many condemning references to the phenomenon of Israeli emigrants (*yordim*) to the United States. For example, "These persons who deserted our country are good people, friends of our own son, who now rests behind the stones in the military cemetery. . . . These stories [of emigration] are especially painful for us, as citizens of the State and particularly as parents of a son who gave his life for the very existence of this State" (1984 bulletin, p. 7). The problematic exclusion of yordim from the Israeli collectivity has been widely discussed elsewhere (notably Shokeid 1988). It is interesting to see how the discourse of bereavement appropriates it as part of its worldview.

Other disputes arose within the bereaved community itself in regard to the government's standardized commemorative practices. The majority of members remained in favor of standardization. Although standing in stark contrast to the rhetoric of uniqueness, this attitude should be understood in the context of collectivism. In reply to demands to individualize the script of MDF ceremonies, one member argued that

> the reading of personal names [of the fallen] is negligible. Today, to our sorrow, ceremonies are attended by fewer and fewer, and this does not honor the fallen. We should therefore limit ourselves to one central, standard MDF ceremony that will include all of the fallen and be conducted in a respectable manner while involving not only the bereaved families but the whole of the Israeli community. The standardization of ceremonies will ensure that the memory of the fallen will be kept forever, just as the memory of the demolished temple has been kept for more than 2000 years. (1990 bulletin, p. 18)

Standardization was also defended in connection with military cemeteries. Requests for individual design (especially of epitaphs) were heard from bereaved families whose sons were killed in training accidents. These parents could not tolerate the standard epitaph "died in the course of duty." In their bulletins, YL members once again assumed the role of the guardians of the "authorized version." "Although civil and military cemeteries are located one near the other," read a representative piece,

> military cemeteries are much prettier, [well] designed, respectable. All the tombstones are standard and modest. In their life, our sons were

different: different in character, in rank, in duties—until they fell in battle. In the face of death, all are equal. It is precisely because of this that we regret to see a few of the bereaved families destroying the order and the standard design and introducing changes, out of good-will of course. These changes give the cemetery an appearance of a "slum." The military cemetery is a public, not a private, domain, and as such it must contain order and standardization. I have seen many military cemeteries around the world and they are all built according to the same orderly principles. (1978 bulletin, p. 60)

For YL members, many of whom depend on the state's financial support, collectivism is a source of political as well as personal power. Although each carries his or her own personal sense of bereavement ("flowing in his blood"), it is the body politic, the homeland, that they work to sanctify. In contrast to the public facade erected in the bulletins, the everyday life of many bereaved parents is fraught with contradiction and uncertainty. Recent years have seen more and more clashes between bereaved parents and the IDF. Parents have begun to question the circumstances of their sons' death more openly, especially in cases of training accidents. There have even been lawsuits by parents that have reached Israel's High Court of Justice over having something other than the standard epitaph inscribed on their sons' tombstones. Moreover, whereas bereavement generally provides a social license to review and change one's life, when the demands of bereaved parents are considered closely, it can be seen that rather than challenging the public discourse, they reappropriate it in ways more beneficial and reward-ing to them. This bottom-up process has been going on, in exactly the same manner, throughout Israeli history—not only in recent years, but also as early as immediately after the 1948 war (see Sivan 1986). However, it has only been in recent years that the very ideology of bereavement has been con-tested, and its national ethos questioned.

Reading through these lines, I feel that something is not right. My own voice would like to interrupt the all-too academic writing that glosses the personal experience of bereavement. My Master's thesis on the life stories of bereaved parents in the October War (Yom Kippur) began in an unexpected way. When the war broke out, I was in my early twenties, a mother of a one-year-old daughter and the wife of a conscripted husband. I remember getting a telephone call from a university colleague who said that she was looking for volunteers to visit and look after bereaved parents. My consent was immediate. She told me I didn't have to do it, and that it was a very difficult experience. I insisted. My daughter, Tami, then in the grip of a high fever and cranky, tried to pull the telephone away, crying into it. My mother was in the living room, and I motioned to her with my free hand asking her to take Tami from me. Mother refused. She did not approve of my voluntary conscription to help the bereaved, rather than my own family.

The next day, I showed up at the Ministry of Defense. I had to leave sick Tami at home, a sickness probably worsened by missing her Daddy AND her Mommy. I left in the morning and came back at night. The next day was the same. And so was the day after that. Every day, another journey. In the morning, arriving at the office, I received a list of bereaved families scheduled for visits that day. I had to take a bus to each. At first, I accompanied an elderly war widow (her husband fell in the Sinai War, around 1956), who used to explain to me that it was more difficult to be a war widow than a bereaved parent. This question would arise again many times. Later, I went out by myself, and I found three bereaved families that I considered "my own," to whom I would return again and again. It was as if I were possessed by the experience, as if I knew this was my place, a place of death that had in it the essence of life. Homes where the walls did not just break down but rather opened up like flower petals. All secrets were in the open. There was no point hiding anything at a time like that. And I was there, feeling naked too, without masks, mythical. I was under their spell. I felt that I had found life in the place of death. I was both terrified and possessed by the disaster. I wanted to see more and more disasters, feeling that some essential truth was there for me to find. They thanked me, but it was I who wanted to thank them. I had then the feeling that they had taught me, more than anyone else, about life. I gave up my own life, I was with my (adopted) bereaved families almost every day, and I even dreamt about their fallen sons. Everything else seemed puny, negligible, next to the disaster.

I have clung to my bereaved families. I lived with them and I have written about them, and thus I betrayed them. I felt that it was a betrayal, to make cultural capital from another person's tragedy. I was cold when I woke up at night thinking, what kind of academic vulture am I. People have opened their homes to me, giving out their secrets, and I have betrayed them. And I was afraid that some kind of terrible punishment would be inflicted on my family—that my daughter would be taken, and someone would write a thesis on me. Gradually I settled down. I was not the only one who wrote about other people's suffering, was I? What about psychologists? Doctors? But calm I was not. I was haunted by fears of becoming bereaved. Thoughts of what if this were to happen to me. What would be more difficult—losing my husband or my children? It was my identification with "my" bereaved families. These were the kind of questions they were asking, questions that separated them from the rest of the world, the "normal" world. I was trying to "feel" into their disaster, to be on their planet, speak their language. A few years later, when a colleague told me about her academic work, which dealt with children's everyday games, such as stamp collecting, and birthdays, I was really surprised: how she could deal with such "banal" things while I was immersed in human tragedies?

At some point, I realized that I had to let go, say goodbye to my bereaved families and move on. It happened when I wrote up the research. Writing provided the necessary academic distance. Again, at first I felt like a traitor and didn't agree to have my thesis put in the library. But I realized that there was also something else in the refusal: a "psychological crime" not merely against the privacy of my bereaved families but also against the Israeli public. My thesis included material about how some of the bereaved parents spoke against the country, questioning its moral integrity. In those years, no one dared publish such stuff. We were still captives of the Zionist dream. My refusal to let the thesis "go public" also reflected my fear of the "eyes of my country."

I defended my thesis when I was nine months pregnant, a few hours before the birth of my son Shay. I remember that no one questioned me how it felt for me to do this research while I was pregnant. Maybe it had to do with the fact that there were only men in the room. No one asked me, either, how it felt to be done with the research.

The Art Exhibits: Bereavement as Counter-Ideology

Two art exhibits mounted in the early 1990's can be usefully discussed here in connection with the bereavement cult: "Tending to Get Caught in the Details" (artist Gili Meizler, Limbus Gallery, Tel Aviv, November–December 1994) and "The Act of Photography in the Confession Cell" (artists Nir Nader and Erez Harodi, Bugrashov Gallery, Tel Aviv, February 1993). Both exhibits featured the works of men who had lost a close relative in battle—a father in the case of Harodi, an older brother in the case of Meizler.

Meizler, in interviews with two newspapers (*Kol Ha'ir*, 18.11.94; *Ma'ariv*, 18.11.94), said that his exhibit was the result of a visit to the tank warriors' memorial in Latroun, where his older brother's name was inscribed. He described various items that caught his surprised attention in the memorial building, such as large rocks labeled "taken from the battlefield in Sinai" (or "from the Galilee," and so on), a computer where one could get a printout on one's fallen relative, and a souvenir shop where one could buy puzzles of tanks, white postcards with the writing "Uzi Does It," and plates, hats, and coffee cups decorated with military badges.

For Meizler, the war memorial demonstrated the dark side of the collective ideology of bereavement. His exhibit was a direct attack on what he perceived as the standardization and commodification of bereavement. "There is not one corner in the Latroun complex where visitors can really cry," Meizler said (*Ma'ariv* interview, p. 18). "After you were photographed on the tank, went through the shop and bought a bottle of wine with the unit's badge, after all this you will not shed a tear in front of the memorial wall."

Meizler's exhibit was made of representative items ("*objets trouvées*") replicated from the memorial (e.g., a printout, an admittance ticket, souvenirs from the shop, photos of other items). In addition, the wall of the exhibit room was covered with an enlarged version (6x9 m) of the complete index of the book *Loss and Bereavement in the Jewish Society in Israel* (Wiztum & Malkinson 1993). It contained such entries as "social insurance," "collective memory," "nightmares," and "loss." His so-called Wall of Names was a sarcastic allusion to a real memorial wall; replacing the names of the fallen with the index items was his way of protesting the standardization and trivialization of death. The individual soldiers, in their capacity as the fallen, are mere index items in the national hall of fame. On entering the exhibit, visitors received a printed excerpt from George Mosse's revised Hebrew version of his book *Fallen Soldiers* (retitled *The Fallen in Battle* and published in 1993).

In his interviews, Meizler referred to Harodi and Nader's 1993 exhibit as a direct influence: "The idea did not come from them, but there was a dialogue, I appreciate political art." Harodi and Nader (H&N, for short), two photographers who were already known before that exhibit for their political, anti-Establishment artistic activity, were indeed influential in introducing the subject of bereavement into Israeli art life. For this purpose, they turned the gallery's largest chamber into a memorial-like hall, with dimmed lights and a row of chairs facing a huge blank wall, bearing the conventional "reserved for the families of heroes" sign used in memorial gatherings. Small poles connected with ropes limited the access to the chairs. A display case with army shoes sprayed in gold (and hanging upside down) stood in the entryway. Prominently placed inside the room itself were two wooden chests reminiscent of army coffins, each containing a Keren Kayemet (Jewish National Fund) "blue box."[10] A carpet of the same color as army uniforms covered the floor. A book accompanying the exhibit bore the subtitle "Making a Living of My Father's Death." In a smaller room, H&N placed a miniature model of the larger chamber. Here the relations of representation were reversed, and the observer gazed at the exhibit instead of being subjected to the exhibit's gaze.

Although the symbolism of H&N's exhibit was multifaceted,[11] I will be concerned here solely with decoding its reference to bereavement. What H&N showed us was arguably an exhibition of protest and resistance. They seemed to offer a counter-ideology of bereavement by exposing, and deconstructing, its major discursive mechanisms. First, H&N were interested in consolidating (and thus problematizing) the boundary between the bereaved families and "the rest of Israel." They created a simulation of a memorial hall where the observer, as a peeping Tom, was forced to distinguish him- or herself from bereavement, to regard bereavement from without, to interpret it as a discourse. In H&N's confession room, bereavement became objectified, and the bereaved subjects.[12] The ropes that limited access to the chairs served as a further symbolic demarcation between "us" and "them," and the "reserved for the families of heroes" sign became decontextualized and therefore ironic. In addition, H&N were evidently interested in laying bare the depersonalization, commodification, and glorification underpinning the ideology of bereavement in Israel (the gold-sprayed army shoes). Their blank wall where there would normally be names embodied this criticism, in a parallel manner to Meizler's indexed wall.

A further dimension in H&N's exhibit was the intended contradiction between the public and the private. The artists placed bereavement, that private, sensitive, unique experience, in a public sphere of normative control. Bereavement, as we have seen, is prescribed in bureaucratic procedures, confined to controlled zones and dates, and glorified to suit the collective ethos of sacrifice and resurrection. The individual meeting with bereavement is always, ipso facto, made under the gaze of the collectivity, in the national panopticon, and on an olive drab carpet.[13]

In 1993, Yad Labanim published a letter in its bulletin from a bereaved father who had watched a television interview with Harodi and Nader in connection with their exhibit, which he later visited. Unlike Segev's attack on the bereavement ceremonies, Harodi's was made "from within," by a member of the bereaved community. Accordingly, the letter writer used a milder, rehabilitative tone in his comments. It is interesting to note that he did not once mention Nir Nader, ignoring, that is to say, a man who had not suffered a direct loss of a family member, as Harodi had. Captioned "To the Sorrows of Bereavement," the letter read:

> He [Harodi] found a cure in attacking those who tried to help him and to encourage him to complete the tasks his father could not complete. However, Harodi is wrong in turning his anger into a mockery of society and its institutions, for example by his cynical use of the army shoes. He said in the interview that he refused to serve in the army, but remarked that he sat in jail only two weeks, because he was once more "making a living of his father's death" since the IDF did not feel comfortable locking up a war orphan. In contrast to Harodi, I have felt committed to the IDF from the moment my son fell.

The writer concluded by suggesting that a more cautious psychological follow-up should be given to war orphans. Harodi's protest was dismissed as a transgression, a psychological dysfunction that could and should have been taken care of during childhood.

My reflections on these exhibitions and on bereavement in general were influenced by my talks with Meir Gal, a curator and artist, who was working on several exhibits on militarism and bereavement. In our talks, he explained how he experiences others' bereavement, how death enters his life through the morning news, while listening to the radio over breakfast, sipping coffee while hearing routine reports on casualties of terrorism, border patrols, Intifada, the Lebanon front—the all-Israeli experience.

Meir explains and interprets the exhibit of his artists. When I speak to the artists, they are less assertive than Meir; their work speaks for them. I see the similarities between the curator and the anthropologist, between Meir and myself. I also explain and interpret "my Artists." Meir Gal reminds me of myself, Meira. But he is more daring, angrier. His own exhibits seek to turn the innocent and taken-for-granted into vulnerable spots. He has left Israel and chosen to live in exile.

Meir tells me: "One of the important things about living abroad is creating a distance between myself and the country [Israel]. Not to be used as a soldier, a victim, a subject." When I tell him about my interviews with the "Children of Yemen," he screams, "Let it all burn in Israel, I wish everything there would burn and be demolished." I feel he has just hit a fundamental element of my Self, and we don't talk again for a very long time.

The National Cult of the Dead

One is haunted by the thought that commemoration is always, and inevitably, a collectivist project. As such, it cannot but reproduce the very conditions that led to the death of young people in the service of their nation in the first place. National ideologies of commemoration have several recurrent motifs, one of the most prominent among them being sacrifice. L. Lloyd Warner (1959: 249) has written, in his classic ethnography of Yankee City, that the principal theme of the Memorial Day ceremonies there was sacrifice: "the sacrifice of the soldier for the living and the obligation of the living to sacrifice their individual purposes for the good of the group." But in fact Israel makes much more of the cult of the fallen than any U.S. city does. In Israel, in addition to the nationwide school programs, each of the underground (prestate) military organizations and various units of the IDF hold annual rites of commemoration at other points in the year; and further regular meetings and ceremonies are conducted in Yad Labanim centers, which exist in every local Israeli community. All the texts mentioned in this chapter portray the profound engagement of Israeli society with commemoration, an engagement so extensive as to be termed a "national obsession" (Weingrod 1995) and a "national cult of memorializing the dead" (Aronoff 1993: 54). Although in the United States, Memorial Day and bereavement have not preoccupied the public (at least not until the events of September 11, 2002), in Israel they have a public and felt presence throughout the year.

The Israeli "obsession" with commemoration, besides being factually grounded in the sheer number and frequency of war and terror casualties, is arguably rooted in two major sources: (1) the political use of commemoration as a symbolic mediator between past and present, and (2) the use of commemoration for social mobility. For the Zionists, the single-most-important problem in their nation-building project was finding a way to construct a historical bridge to a land from which the Jewish people had been exiled for nearly two millennia. "Perhaps the primary goal of Israeli political culture," argues Aronoff (1993: 48), "has been to make the continuity of the ancient past with the contemporary context a taken-for-granted reality." This obviously became all the more important given the challenges to the Jews' right to statehood, by Jews as well as Arabs. The fallen thus became visible, unquestioned evidence for this right to statehood. Furthermore, the sacrifice of the fallen—often identified in American culture with the sacred life sacrifice of Jesus Christ (Warner 1959: 279)—was connected in Israel with ancient biblical heroes, as well as newer heroes among the pioneers.

Two stories often mentioned in conjunction with the commemoration of Israeli soldiers are those of Masada and David and Goliath. The Masada tale, which has become for Israelis, an ultimate myth of Jewish sacrifice in the face of a superior enemy, is told so that "Masada shall not fall again" (Schwartz et al. 1986). The induction ceremony for new army recruits, who grasp the Bible in one hand and a gun in the other, was traditionally held at Masada or at the Wailing Wall in Jerusalem. The biblical story of David and Goliath, again an account of the few versus the many, is likewise used to mobilize citizens into a state of perpetual conscription and a feeling of siege (Gertz 1984). Also often summoned up is the legendary battle at Tel-Hai, where six Jewish settlers died on March 1, 1920, while defending a small northern settlement against Arab forces. The reputed dying statement of Yoseph Trumpeldor, leader of Tel-Hai, was "Never mind, it is good to die for our country" (see Zerubavel 1990, 1991, for an analysis of the Tel-Hai narrative). Tel-Hai Day has become institutionalized as an official part of the Israeli culture of commemoration. All these commemorative stories reinforce the Israelis' sense of themselves as the few against the many, a sentiment that has become one of the most central in the Israeli culture of the chosen body.

As for the other political usage of the cult of the dead, as previously mentioned, in Israel the military is arguably the most important social network. Despite the events of the last few decades—especially the war in Lebanon and the Intifada—for most Israeli men, participation in the army is still considered to be a reward in itself. The fact that, in earlier times, kibbutz members disproportionately suffered high casualties in front-line units was often cited as evidence of their vanguard role in society. The same sentiment has been more recently expressed by other groups. Myron Aronoff (1989: 132) reported that in interviews conducted during the war in Lebanon, leaders of nationalist religious Jews and Sephardi Jews told him that the disproportionately high rates of casualties suffered by their respective groups were "evidence of their having moved to the forefront of the national struggle." The death of family members in military service, as a closer ethnography of individual bereaved families should illustrate, is also used as a means for personal gains in the form of financial benefits and a social license to change one's course of life (see Weiss 1978, 1989). The art exhibits described here are in themselves a means of professional mobility—"making a living of my father's death."

The political use of commemoration is not limited to the context of bereavement. A particularly important example of its use as a lever for social mobility is the recent upsurge of new shrines for Moroccan saints (zaddikim) whose bones were transferred to various outlying Israeli towns where the majority of the population is of Moroccan origin. Members of this, and other North African ethnic groups, have come to take part in the Israeli cult of the dead by organizing large pilgrimages to the new "memorials" (Weingrod 1995; Bilu & Ben-Ari 1992), and thereby to stake a claim for "legitimacy and status as equal Israelis" (Weingrod 1995: 16).

The Israeli culture of commemoration, or cult of the dead, therefore has a dual ethnic nature. On the one hand, it strives for a collective body—an archetypal sabra—cleansed of individual features such as ethnic origin, gender, age, and rank. It ultimately presents itself as a key symbol cutting across historical periods and ethnic divisions. In that sense, it belongs to what has been called "the language of commitment," a language characterizing communities that are "in an important sense constituted by their past" (Bellah et al. 1985: 152–53). These so-called "communities of memory . . . carry a context of meaning [that] turns us towards the future" (Middleton & Edwards 1990: 5). This language of community commitment dominated Israeli society until the 1980's, when the opposite language of the self-reliant individual began to challenge its supremacy (Eisenstadt 1985; Lissak & Horowitz 1989). As the Israeli routine of military conflict (see Kimmerling 1974, 1985) created both collectivism and bereaved families, the two became interdependent. The collectivity glorified its fallen soldiers and financed their bereaved relatives, and bereaved families committed themselves to the collective ethos of sacrifice and the standardization of bereavement.

It is important to stress, in this connection, that the chosen body being commemorated is above all a Jewish body. As we have seen, traditional Jewish religious practices prescribe the burial of non-Jewish soldiers outside the cemetery's fence. This literal exclusion is exercised where there is doubt of the fallen soldier's Jewish identity, a fate that has befallen many immigrants from the former Soviet Union. The criticism of a liberal minority ("this soldier was good enough to die in the service of Israel, but not good enough to be buried as an Israeli?") has so far had no effect. Interesting in that connection is the reaction of the family of a fallen soldier from Russia who was recently buried "outside the fence" because his mother was not Jewish. They "did not mind," they told reporters. "For us he is Jewish irrespective of where he was born" (*Yediot Acharonot*, 22.2.02, p. 6). The persistence of this traditional practice marks a nondiscursive continuation between the Israeli and Jewish identity. I call it nondiscursive because it seems to hinge on some primordial, atavistic frame of mind that sometimes seems to defy rationality. The monopoly of the religious establishment in the burial and undertaking market, for example, is indisputably taken for granted. Any secular criticism of the Chevra Kadisha's rigid adherence to the prescriptions of the Jewish code is considered profane by many Israelis and dismissed as questioning the very foundations of their existence.

Handicapped Bodies and Sacrificial Bodies in Israel and Traditional Judaism

My aim in this section is to link the previous chapters together by examining the ways in which the body connects the new Israeli identity to the traditional Jewish collective identity and the ways it separates them. Israeli identity, it is often claimed, was constructed out of the rejection of traditional Jewish identity, especially as it related to the diaspora. Nevertheless, the state of Israel was established in accordance with the traditional Jewish code of conduct. In practice, the halacha still prescribes major events in the life of Israeli Jewish citizens. However secular the everyday life of many Israelis may be, they are still subject to the rule of the halacha in their rites of passage—all of which have carnal aspects—such as birth (the circumcision ritual, *brit*), marriage, burial, and bereavement. Body impairments, like the body of the deceased, are a major issue in traditional Judaism and are thoroughly treated in the halacha. In this section, I compare the traditional Jewish articulations of body regulation in terms of impairment and sacrifice with their contemporary counterparts in Israeli society. My overall argument will be the underlying continuities between the two identities. I argue that Israeli and Jewish identities, although sometimes discursively (politically) separated, are closely tied in more fundamental, nonverbal practices—for example, those related to the body.

Handicapped Bodies

What does it mean to live under the spell of the chosen body? If the collectivity defines its subjects through their bodies, this must involve a pervasive process of screening. In a nation in arms and an army of labor, the desired bodies are those of soldiers and workers. People unfit for such national service are bound to be deemed marginal. And that is exactly the case in Israel.

According to the report of the Public Committee on Legislation Concerning People with Disabilities, the approximately 10 percent of the Israeli population who have disabilities are a "minority group that suffers discrimination, segregation, and stigmatization" (1997: 12). The report paints a bleak picture in which people with disabilities are ignored by most of the nation's institutional systems. For a start, Israel has no comparable law to the Americans with Disabilities Act (1990), the Canadian Charter of Rights and Freedoms (1994), the British Disability Discrimination Act (1995), or the Swedish Act Concerning Support and Services for Persons with Certain Functional Impairments (1993). Moreover, the nation's equal employment opportunity law, while prohibiting discrimination on the grounds of gender, nationality, religion, and so on, does not mention disability.[14] Many public buildings are inaccessible to the handicapped. Most of the people with developmental disabilities live in institutions (approximately 80%). There is no national agenda for sponsoring or budgeting for leisure activities for the disabled. The committee recommended legislation that would attend to all these matters. In sum, the rights of people with disabilities are neither protected by law nor accommodated in practice.

Casualties of war pose a particularly knotty problem, not only because of their relatively large numbers, but precisely because they have given part of their bodies for their country. In principle, handicapped veterans should receive as much symbolic acceptance and gratitude as their fallen comrades. But according to the commission, they do not. Although handicapped veterans receive ample financial support from the Ministry of Defense's Rehabilitation Department, in practice they are segregated. There is no symbolic difference between them and ordinary citizens with handicaps. Both are similarly labeled by the larger society. Israeli society, as a whole, is inaccessible—physically and mentally—to both. No accommodation is made for them in public schools and, astonishingly, hospitals, let alone museums, clubs, swimming pools, cinemas, and the like. Public transportation is not equipped for these people either, but at least in this case, handicapped veterans are entitled to specially equipped cars. Symbolically, however, the "specially equipped car" is also a mechanism of seclusion. "We [Israeli society] are practically turning our backs on [the disabled]. You are handicapped? Don't go to a restaurant—eat a sandwich at home" (Michael Shiloh, the Israeli ambassador to Norway, in a letter cited in the committee report 1997: 42).

In his testimony before the committee, Doron Yehuda, chairman of the Coalition of the Disabled in Israel, an organization representing more than 500,000 people, leveled similar accusations regarding the persistent discrimination against them. "We can't get jobs because we can't get into most buildings," he said. "If I do find a job, I can't use the toilets because they're unsuitable." In 1997, when terrifying footage of the helicopter crash in She'ar Yashuv appeared on TV screens, the country's deaf population was thrown into a panic, Aharon Eini, chairman of the Association of the Deaf in Israel, told the committee. Among the country's 300,000 hearing-impaired people were parents of soldiers serving in Lebanon, but without subtitles or signing, they could make no sense of the bloody, fiery scene. The symbolic exclusion and segregation of people with disabilities—and particularly of handicapped veterans—is thus another, and particularly harsh, testament, to the cultural hegemony of the chosen body. Once disabled, a person cannot in any way be a chosen body.

These policies echo the traditional halachic view that the covenant necessitated an able-bodied nation. My discussion here is guided by the need to locate these practices within the broader traditional context of Judaism and the halacha, as it relates to the body. I begin with the handicapped body, asking whether its view in the halacha can be seen to enable or even justify the Israeli discourse of the chosen body. Rabbi Tzvi Marx, whose manuscript on the subject I consulted, claims that halachic views of the disabled are filled with tension and ambiguity. There are halachic guidelines that define the whole person by his handicap and thereby recategorize him as functioning outside of ordinary norms. A premise of Judaism is that because man is made in God's image (*tzelem Elo-him*), he is endowed with a special entitlement in comparison with the rest of the created order. The prohibition against the murder of non-Jews or the failure to promote birth was construed, besides all else, as an affront against God whose image man reflects. Man's special entitlement consists most broadly in human dignity but includes his cognitive and speaking abilities, as well as his capacity for freedom and spontaneity. So by halacha definition, "those that lacked these capacities, like certain of the handicapped," were automatically excluded (Marx 1993: 159).

In general, then, traditional Judaism saw disability as a stigma. This is evident, for example, from the disqualification of a blemished priest from serving in the temple and the disqualification of blemished animals from sacrifices (Marx 1993: 296). Jewish priests (*kohanim*) were to be disqualified from the important and public ritual of "priestly benedictions" (*birkot kohanim*) not just for serious moral lapses, but also for physical deformities. As the Mishna (Meg. 4:7) taught, "a priest whose hands are deformed should not lift up his hands [to say the priestly blessing] and [Rabbi Judah, the compiler of the Mishna] says: Also one whose hands are discolored with a red dye should not lift up his hands because [this makes the congregation look at him]."

The halachic culture is described as one that celebrates competence (Marx 1993: 680). The covenant at Mount Sinai, so the argument ran, necessitated competent partnership in the form of an able-bodied nation. To underscore this point, a rabbinical commentator labored to prove that those who stood at Sinai were fully recovered from their disabilities, rendering them capable of receiving the Torah with "the dignity of whole-bodiedness":

Whence do we know that there were none lame among them? For it says, "and they stood at the nether part of the mountain" and stood can only mean upon one's feet.

Whence that there were none with broken arms among them? For it says, "All that the Lord hath spoken we will do."

Whence that there were none deaf among them? For it says, "And we will hear."

Whence that there were none blind among them? For it says, "And all the people saw the thunderings."

Whence that there were none dumb among them? For it says, "And all the people answered." (Cited in Marx 1993: 680)

It is therefore by whole-bodiedness and able-bodiedness that the norms for complete covenental participation were set. The nation must be capable of carrying out the halachic imperatives, the *mitzvot*, to the exact prescribed detail. It is only within this given, says Marx (p. 681), that "we can inquire as to the halacha's flexibility, whether it is able to bend its demands to include (by obligation or otherwise) even those less minimally competent according to its initial directives."

A concise overview of the halacha, in sum, indicates that (1) it provides a basis for a normative discourse of control regarding the body, and (2) sees handicapped bodies as unequal to whole-bodiedness and able-bodiedness. So Israel's indifference to the plight of the handicapped stems, in the first instance, from the halachic approach to the handicapped body. But that approach was reinforced by the revulsion toward the Jews of the diaspora that was so deeply a part of the Zionists' ideology, and that led to their determination to invest the renewed fledgling Israeli society with the "perfect" man (see Fischer 1988 on Zionism as a salvational realization of Judaism). Handicap is, on this view, a reminder of the Jew's "crippled" condition in pre-Israel times, undermining the dreams, the exaggerated visions of naive Israeli ideology, and is therefore rejected as counterproductive to the enterprise of rebirth.

Sacrificial Bodies

At several points in the discussion, I have mentioned that sacrifice, in particular the traditional Jewish concept of sacrifice (*akeda*) as embodied in the fallen soldier, is a central theme in Israel's culture of the chosen body and a crucial element in the Israeli ethos of bereavement and commemoration. It is time to ask where these themes of idealized sacrifice and the purity of the fallen originated.

In a commentary headlined "Ripples of a Bomb" (*Jerusalem Post*, 4.2.96, p. 6), Michael Cohen, president of the Reconstructionist Rabbinical Association of the United States, compared Israel's (and his own) loss of sons in terrorist bombings to Abraham's sacrifice. He opened his piece with a quotation from a poem by Yehuda Amichai: "The diameter of the bomb was thirty centimeters and the diameter of its effective range about seven meters." Effortlessly, Amichai then goes on to describe how the effect of the bomb rippled to "the distant shores of a country far across the sea," until it "includes the entire world in a circle." Within that circle, Cohen wrote, he had recently and painfully discovered his son, not yet four years old. "Up until the recent reign of terror in Israel, Roi was so excited about us going to Israel next year when I will be on sabbatical leave from my position as a rabbi in Vermont. . . . He and I talked about how on Shabbat we would walk together to synagogue on the streets of Jerusalem, and that he would be able to sit with me during services since I would not have to lead them. Then came the bombs and the . . . buses turned into crematoriums on wheels in the heart of the Jewish nation. I think: Am I Abraham bringing my son to Mount Moriya?"

God commanded Abraham: "Take now your son, your only son, whom you love, even Isaac, and get you into the land of Moriya, and offer him there for a burnt-offering." If one had to choose a moment in the biblical account that evokes awe, confusion, inspiration, and terror all at once, Abraham's binding of Isaac, the akeda, stands out. For thousands of years, commentators have grappled with this unsettling event. Abraham had been promised that a nation emerging from his own loins would, against all odds, survive and declare the name of the One God to the world. Moreover, Abraham's message to the world was one of ethical monotheism—of a God who deplored immorality and bloodshed. The consensus verdict of those commentaries is that by his willingness to subjugate his own will to God's, Abraham proves to be the ever-faithful servant, whose immense faith is rewarded. Most commentators approach the akeda from Abraham's point of view. Maimonides, for example, saw it as depicting a fundamental truth of prophecy. Abraham does what he does because he has heard the Divine Command; a true prophet cannot but obey. Thinkers like Soren Kierkegaard and, closer to our own age, Yeshayahu Leibowitz understand that Abraham is being taught an essential lesson of religious worship—that of suspending one's own judgment. Jews historically, and especially during the Crusades and other times of persecution, have looked to Abraham for inspiration. The *Zikhronot* prayers on Rosh Hashana, in asking for divine forgiveness, speak of the myriad of Abrahams who have evinced a similar commitment (*mesirat nefesh*).

In modern Israel, the biblical legacy of Abraham serves as a paradigm for times when Israelis must risk their lives and the lives of their children for the preservation of their faith and people. The sacrifice provides a cultural lens through which Israeli existence is perceived. It is a very selective adoption of Jewish symbolism by the contemporary Israeli establishment. As we saw, the Memorial Day ceremony for the fallen almost always culminates in the reciting of the poem about the young soldiers who sacrificed their lives as the "silver platter" on which the state was served. By virtue of falling in battle, these sabra-soldiers became the chosen sabras. The motif of chosen-ness, of singularity, is here transferred from the religious to the nationalist realm. Chosen-ness becomes a master narrative glorifying the sacrifice of life in the service of the nation. This militaristic rhetoric later found its way back to religious circles, when ultra-orthodox Jews began justifying the exemption of *yeshiva-bucher* from the IDF on the grounds that "they kill themselves in the tent of Torah."

A perfect body is one feature of the chosen-ness of the sacrificial offering. The perfect body motif is probably not unique to Judaism, but the Jewish culture did seemingly provide a uniquely negative approach toward bodily excreta and impairments, as Mary Douglas (1966: 124) so graphically pointed out. With the return to Zion, the expectation was that the Israeli body would be repositioned on the stage of history, regaining its health and safety. But in effect the situation has remained unchanged, since Israeli society faces an existential threat on a regular basis, in a similar (or even worse) manner to the Jewish condition in the diaspora. The canonical texts of the IDF are meant, at least in part, to disguise the "pollution" of death while acknowledging the sacrifice of the chosen body.

4 ENGENDERING THE CHOSEN BODY

WOMEN AND SOLDIERING

THIS CHAPTER, on women and the chosen body, takes as its critical point of departure the Zionist thinker Max Nordau, for it was he who more than any other early thinker established mainstream (or as feminists call it malestream) Zionist ideology. Nordau's Zionism was "Judaism with muscles." For him, the diaspora Jew was weak, degenerate, hysterical, and ultimately *feminine.* He wanted nothing less than for Jews to "renew the link to [their] ancient traditions: Be once again men with strong chests, steady limbs, sharp-eyed."[1] This malestream ideology of masculinization was supported by most subsequent Zionist leaders. For example, in a letter written in 1918, a prominent socialist Zionist, Me'ir Yari, declared:

> We want to educate a stiff generation, not people who dwell on dreams.
> Our youth should prepare himself for a life of labor and know his role
> in the land of his ancestors. Everywhere there are rocks and sand and
> desert. Only an arm of Hercules could put this toil into action, and not
> prophets nor poets, . . . not pens, paper, and ink, not poems or hymns,
> not confessions—but axes, shovels, hammers and first and foremost,
> hands! Let us have hands! (Cited in Gluzman 1995: 5)

My concern in the pages that follow is how Israeli women reacted to the view of the "chosen body" as an exclusively masculine one. What was women's place in such a gendered, national, and militaristic discourse? Although the Zionist revolution openly claimed that women were to enjoy an equal role to that of men, the culture of the chosen body in fact had an inevitable effect of gender differentiation (and discrimination). Women—the new Hebraic women—had no other choice but to conform to this manly ethos, but they could do so with a "voice of their own." Even as women reaffirmed the manly and nationalist Zionist discourse, they devised various strategies for subverting that discourse. In what follows, I lend an ear to the voice of the new Hebraic woman. The chapter is divided into three parts. All of them aim to "undo" the voices of women from the malestream narrative of the chosen body. In the first part, I describe the "carnal" poetry of women in the War of Independence. The second part returns to the interviews I conducted with parents of soldiers, this time offering a gendered reading of them. The third part offers an ethnographic analysis of a specific group of women who set out to appropriate the male discourse of soldiering.

The Chosen Body in Women's Literature

Many scholars have studied the ideological role of Hebrew literature in the construction of the new Hebraic man. But as Michael Gluzman (1995) suggests, important gendered aspects of that ideology—of normative masculinity and the repression of femininity—were not subjected to any systematic study. According to Gluzman:

> The construction of Zionist masculinity is such a prominent topic in Hebrew literature that one could actually read the whole history of Hebrew literature in this century—from Bialik's famous story *Arye Ba'al Guf* to David Grossman's *The Book of Internal Grammar*—as a consistent and almost obsessive concern with what is proper masculinity, as well as a testimony to the Israeli culture's attempt to repress the feminine. . . . Tackling the literary project of Zionist masculinity may broaden, in my view, the theoretical discourse of the new historians, whose criticism of Zionism often totally ignores questions of gender and sexuality. (1995: 6)

One study that does focus on the project of Zionist masculinity and its effects on women writers is Chanan Hever's (1995) excellent article on the poetry of the national body. The War of Independence generated a burst of literary works by the first generation of sabras, the so-called Tashach Generation (Tashach is 1948 in the Jewish calendar). It is a generational literature replete with the signs and metaphors of nation-building, masculinity, and commitment (see Yafe 1989; and Miron 1992).

According to Hever, women poets had a considerable role in this literature. The poems of Yocheved Bat-Miryam, Haya Vered, Anda Amir-Pinkerfeld, and Edna Korenfeld appeared regularly in literary magazines, in the IDF journal, and in poetry collections that were published after the war. But widespread publication did not raise these women's status as poets, and they remained in the margins of the literary canon.

As in other segments of society, the women poets committed themselves to the national effort, contributing to the construction and fortification of the Israeli collectivity. They were nevertheless unable to take a full part in it. Since war was perceived as a masculine affair, its experience could be fully captured only by those actually in the front lines, which is to say, men. This was true despite the heroic image of women warriors (whose activity was discontinued in the first month of the war; see Sivan 1991: 3–39). Although the palmach conscripted women as fighters, and the Labor movement championed the equality of women, the overwhelming majority of women were relegated to service functions in the rear. The IDF established the Women's Corps (CHEN, short for Chel Nashim, Females' Corps) in 1948 as an administrative cadre governing the training assignments and military careers of women. (It is no coincidence that the Hebrew word *chen* means "charm"; in the words of an IDF spokesman, "CHEN adds to the IDF the grace and charm which makes it also a medium for humanitarian and social activities"; cited in Yuval-Davis 1985: 663.)

The male contributors to this corpus of war poetry, all or most of them soldiers, emphasized the symbolic immortality of the fallen through the central image of "the living-dead" (Miron 1992). In many poems, the fallen are brought back to life and address the nation. Alterman's poem "The Silver Platter," a sure candidate for almost every national rite of commemoration, is a prominent example, almost a prototype of the genre, and I have already discussed it in Chapter Three. This mythical image of "the living-dead" was an ideological mechanism reaffirming the readiness to sacrifice personal lives for the sake of the collectivity. The living-dead is a fallen soldier whose personal body is missing, but whose imagined, national presence is high-lighted.

The women poets used that image too, thereby affirming their commit-ment to the nation. But their usage of this malestream image was unique. They appropriated the image and located it within their own female dis-course, speaking about the fallen from the position of mourning mothers and lovers. Even as they accepted the need for the sacrifice to the collectivi-ty, they often also emphasized the metonymy of the flesh, the personal, irrecoverable death. Hever (1995: 111) cites the following poem by Bat-Sheva Altshuler (1949), titled "He Was a Tank Warrior" *(Hu Haya Tankist)*:

> Only one more was left
> Lying in the shaft
> A body stamped in flesh.
> One of the black men took care to
> take off his shoes.
> The body black, spotted with blood
> Surrounding, the colorless silence.
> Where is the gold of wheat?
>
> Where is the golden-fire of warriors?
> From a shaft, he has risen,
> A bloody voice of mourning.
> He has fallen with the fallen,
> And he was a tank warrior.

As discussed in the chapter on national commemoration, the masculine discourse of the fallen soldier is first and foremost a sterile discourse. The fallen, in Alterman's "The Silver Platter," rise from the dead to speak to the people; they have fallen in battle, yet their bodies are "dripping the dew of youth." The living-dead, in the men's poetry and literature, is typically beau-tiful, handsome, ideally remembered. He speaks "to the people." His voice, words, and body are a public construction, spoken, configured, and realized in public. The same sterility characterizes the works of such renowned men poets as Ayin Hilel and Chaim Guri. Guri, for example, writes about "the silent 'Mountain Squad' [whose members] would flourish once the last shooting cry has faded in the mountains." The fallen soldiers are here poetically reborn as flowers. Ayin Hilel speaks of the fallen soldier as "returning to your mother . . . the soil." His poem describes how the body of the fallen enters the soil, fertilizes it, becomes a part of its rhythm. In the poems of the "chosen body," the soil is the real mother of the soldier.

Against the sterile distance and nationalistic glorification of the chosen body by men poets, the women poets emphasized the private side of bereavement through the filthiness and bloodiness of death. Against the abstract image of the living-dead, women poets spoke plainly about specific persons.[2] Against the bodiless voices of romanticized heroes, these women wrote about the need to touch the body: "My son has fallen in a distant place—and my hands did not have the privilege of touching him. Here it is a great privilege, touching the dead body of the son before his burial, embrace him, wash him with tears" (a bereaved mother of a 1948 soldier, cited in Hever 1995: 115). These very personal comments bring to mind the mourning poem printed in the Yad Labanim bulletin. As in the case of the professional women poets, it was the bereaved mother who—in the midst of the authoritative discourse of commemoration inculcated by the Yad Labanim organization—felt free to emphasize the unique bodily aspects of bereavement: "I am no longer like you/I am something from a different planet."

According to Hever, the appropriation of the fallen's body became a source of power for the women poets: "Through the body, which is missing in the context of the masculine subject, the woman generates authority that is born out of the contact with the flesh, a contact that is unique to the woman-mother. Against the motherly contact with the flesh, man generates war, the touch of death, and the image of the living-dead soldier." But there is a second part to this appropriation, he continues:

The women's private contact with the flesh is presented to the public: lines of fallen bodies showcased in poems. This transformation reverses the traditional power relations: Now the woman-mother is the beholder, she is gazing, she has the power to describe and to highlight. It is the woman who spells out, in the public realm, what has traditionally been hers only privately. She gives the unspoken bodily contact a public, discursive representation. Only now, after her son or husband is dead, can she finally represent them in her own voice. (1995: 122)

*Chanan Hever speaks about the combination of personal death and nation-
al birth in the war poetry of men and women. He describes it as a male hegemonic narrative that appropriates the feminine power of birth in favor of the
nation. Men control death (war), therefore men control birth (the nation). The
bereaved mothers whose voice is heard in my research have turned against this
narrative. They resist the national narrative by telling their own story of
bereavement, in their own words. Yael Eyal, a bereaved mother, told me how it
took her nine months to give birth and nine months to put her fallen son "back
in—inside her womb." "This way," she told me, "he is with me all the time.
Under the skin. Like a constant pregnancy." Other bereaved mothers resisted
the national narrative by minimizing its role. According to one, her son did not
go to war; he "went to be with his friends" (who happened to be at the front).
"He was not a hero, he wanted to be with his friends"; and when she thinks of
him aloud, she recalls his body, what he liked to eat, how she made his favorite
dishes for him. These are things that bereaved fathers never spoke of.*

The following section takes us from the poetic and the public to the mun-
dane and the private, as I return to my interviews with parents of combat soldiers. Those interviews show how the discourse of soldiering subjugates both
parents while producing further gender differentiation between them.[3]

My Son the Combat Soldier: Gendered Accounts
of the Chosen Body

In Chapter Two, I described in detail the hegemony of the chosen body as it
appears in interviews with soldiers' parents (mothers and fathers alike).
Looked at carefully, the interviews show that while mothers subscribe to the
ethos of soldiering, they concern themselves more with the personal body
of their sons. Overall, the fathers related more to general societal values such
as commitment, patriotism, and masculinity. Fathers often regarded their
sons' ordeals more stoically, connecting them to their own military experience and explaining them as part of a generalized and unavoidable social
track, perhaps a necessary rite of initiation into Israeli society. Mothers
expressed more concern about their sons' personal difficulties and how they
had to tackle a system that was often insensitive to personal idiosyncrasies.
The following excerpts will illustrate these differences.

Father of Shmuel: "During Shmuel's active military service, I had three other kids in Lebanon [as soldiers], and people asked me how I could be so calm about it. I said that had they been at home, then I would worry. But now they are guarding me, so what do I have to worry about? They are good soldiers. My approach, in principle, is that for everyone who loves this land and sees the state as a value, fear is a secondary thing, and the main thing is to want to do, to contribute. One's personal feelings are not the main thing, and fear can be overcome. I have been through a lot in my lifetime, in the military and outside of it, and it only strengthens my conviction that we must have a strong army, and we need something to protect us."

Father of Itay: "I see it differently [from Itay's mother]. Here, they take your child when he's eighteen, take him for three years—it's part of the country's life-style. What would the guys talk about without the military? But you must also look at it practically—what you do today in the army has implications 35 years after that, until you're fifty. You need to take the career aspect into account. First of all, reserve duties. You do the reserves until you're fifty. If you get out of the army when you're twenty-one, that leaves you with about 30 years of reserve duty. When you're young, the drill can be fun, but later, it's not fun anymore, but an actual pain in the ass. So one should choose his unit carefully because that's where he's going to be stuck."

Mothers rarely spoke in such a tone of rationalization "from above." Their justification often came "from below," from the point of view of their own sons, rather than from the point of view of a generalized other (the citizen, the career man, or the practical adult man). For all that they regarded the idea of "a strong army" as an accepted truth that need not be reuttered, their concern was with the strength of their sons. They were thus both proud and worried at the same time, and their anxiety often centered on the body.

Mother of Uri: "When Uri was conscripted, first he underwent several training courses [*gibushim*] in the navy commando,[4] one of the IDF's top combat units. He told me it was so difficult, he would come home always tired, always bruised; he didn't tell me things, just that it was very secret. Once he was hospitalized for a whole week. I felt great pain because of what he was going through. I wanted to help him out, to be with him when he trained. He got a fracture but continued just as well. On the weekend, I would stand at the window waiting for him, and then I had to wash his clothes so that everything would be dry for Sunday morning, and make him his favorite dishes. I was so proud of him when he made it through the second stage, even though it was so difficult. Toward the end of the course, he said that he had had enough. I was very disappointed, but on the other hand I also felt relief. I was disappointed that he could not persist in what he had wanted to begin with; he had only three more months to go—I was ambivalent."

Mothers, then, allowed themselves a greater degree of freedom than fathers in expressing personal feelings and in admitting that they felt ambivalence. Their discourse leaned more toward the emotional. Fathers were either unfamiliar with such emotions or reluctant to express them. Father of Shmuel: "In basic training, Shmulik and the others were the first cohort of their battalion. When he had the luck to come home, he was really falling asleep on the floor—on the carpet, everywhere. When it was morning and we woke him up, he jumped and called, 'Yes, Sir!' I did not worry. There is a difference between torturing and making life difficult. I knew that Shmulik was strong and could take it. One time in camp, they brought him a letter, and the lieutenant asked him how many push-ups he would do for that letter, so he said 'Five,' so they tell him, 'No way,' so he goes, 'Ten,' and they tell him, 'For you it must be more. Fifty!' And he bends down and starts doing [the push-ups] slowly, and when he reached 50, he suddenly picked up speed and did many more. The lieutenant must have thought he had gone crazy. [Laughing] Yes, he was really in shape."

Overall, it was easier for fathers to look at the body from the outside, generalizing it, (dis)regarding it as subject to greater powers. Fathers therefore more readily took part in the objectification of the body. This approach was presumably a legacy of their own socialization: their own body had been similarly selected, and many of them conducted screenings as military commanders. The cult of the chosen body is obviously more intimately shared by those who have been subjected to it, and the military is a meaningful generational link in Israeli male culture. This disengagement from the personal aspect of body infringement was also apparent in the case of training accidents. When the subject of accidental deaths came up, parents usually differed in their expressed reactions. Fathers often accepted the fatalities and rationalized their acceptance in terms of probabilities. Mothers, in contrast, were often unwilling to "accept" such incidents as fate and dwelled much more on their emotional reactions to such fatalities and the ways they affected their mind-set.

Consider, for example, the father and mother of Itay. Father: "Every day there are training accidents. You open the radio and hear about it. So what? There are more traffic accidents. How many did you say died in the helicopter crash? 70?" Mother: "I can't believe it! Are you trying to say that these accidents do not move you, do not make you think 'that's exactly what is going to happen to my son'?" Father: "No, absolutely not." Mother: "Oh, "I don't believe it!"

The mothers' different reactions to the chosen body open up the old question of nature versus nurture. What is the source of their different point of view? Is it "maternal thinking" (Ruddick 1989) grounded in women's biology, or the result of different socialization? The interviews provide us with a mixed answer.

Pregnancy and birth, two biological events that shape "maternal thinking," did in fact often come up in the interviews with bereaved mothers. The motif of death as birth is elaborated to a considerable extent in a note the bereaved mother Yael Eyal wrote to her dead son:

Birth and death are so much alike: You were born easily and died quickly, the same way. After you were killed in the explosion, I began a count. After you were born, I also started counting the time, noticing the passage of the first week, the first month, the second and third month—Now I count again: the first week, the first month. I have to learn how to be a mother to a dead child. You are still my son. I am trying to make you smaller in my imagination, as small as a fetus. You can be in my womb once again, this is how I can carry you with me.

This motif is also mentioned by Hever, who writes—somewhat obliquely—about the intimate link of mother and son, and how she symbolically gave birth to him anew through the poems. The bereaved mothers I interviewed spoke plainly about their fallen sons as an extension of their own bodies, "Something that you still feel even though the limb is dead [a phantom pain]."

The women's attitude can be accounted for by their socialization as "participant observers" of the cult of the chosen body. I use this anthropological image of identity in an attempt to capture the inherent ambivalence in women's position. On the one hand, they participate in the normative discourse of the chosen body. On the other hand, they observe it from a distance, with a critical and subjective eye, feeling intimately connected to the body they gave birth to but at the same time estranged from the bodily practices that select and mortify it, practices that are made by men and imposed on men's bodies. The difference therefore is in socialization, or its lack. The next section illustrates one instance of women who wanted to be soldiers, but were stopped by (male) society.

The Chosen Body in the Gulf War

During the Gulf War of 1991, many Israeli nurses found themselves caught up in a paradoxical situation. Whenever the attack sirens sounded and citizens all over the country hurried home to lock themselves in their sealed rooms, the nurses drove to the hospital. There they waited for possible mass casualties from gas attacks (which fortunately never took place). In a war unlike any other in Israel's history, these nurses, and not their men, went to "the front," many of them leaving behind husbands and children. A split was therefore created between these women's identities as mothers and nurses. As a national and professional commitment replaced the more traditional, domestic one, that split gave rise to a new rhetoric of soldiering, previously reserved for men only. By focusing on the story of the Gulf War nurses, this section sets out to highlight a real-life example of women's appropriation of the masculine discourse of soldiering, and Israeli society's reactions to it.

Israel's Experience of the Gulf War

For the majority of Israelis, the experience of war is uniting; it is in many cases the formative common experience that links Israelis of different origins, backgrounds, social standing, and religious outlook. Even the Gulf War, which was far from a heroic and uplifting event for Israelis, reflected this tendency toward unification and solidarity. During the Gulf War, daily passivity was accounted for by images from the collective memory: "In this war . . . you are stuck in your own home and you must wait, wait. Suddenly you recall experiences you have long managed to forget, experiences belonging to our personal and collective past in Europe in the 40's" (Stern 1992: 53). When many of the inhabitants of Tel Aviv fled the city after the first missile attack, they were publicly condemned by politicians and others. This was again part of the collective script of the right behavior in war. The "refugees" were blamed for turning their backs on their nation. The popular war slogan of those days was "I stayed in Tel Aviv."

The Gulf War has since been described as the war that turned "the rear into the front" (Danet et al. 1993; A. Ben-David & Lavee 1992). But the war meant different things to different people. For men, it exposed the impotence of Israel and their own; instead of being conscripted and confronting the enemy, they had to passively wait inside their homes. Israeli women were already used to the experience of "being stuck in your home and waiting, waiting" during wars. It was arguably that masculine "shame" that led to the media's systematic deleting of the war almost as soon as the coalition forces left Iraq. The cultural amnesia regarding the war efforts of the nurses was arguably also part of that specific masculine (non)discourse of the war.

The "home front" metaphor had other meanings, however, for Israeli women. For them, it meant a new responsibility for war preparations, a responsibility usually reserved for men. For nurses, it was the hospital that became a "second front," where they went during the crisis, leaving their families behind.

Staging the Gulf War

I conducted open, one-on-one interviews with 15 nurses, three times: in December 1990 (before the outbreak of the Gulf War), in January 1991 (as it was taking place), and in April 1991 (after it was over). The interviews lasted about an hour and were conducted and taped in the nurses' room inside the ward. In addition, I observed the nurses at work in a high-risk zone (Tel Aviv) during the war.

In the postwar interviews, the nurses themselves divided their experiences into those same three periods: before, during, and after the war. Most characterized the prewar period as a time of rumor and panic, including the rumor that Saddam Hussein would carry out his threat to launch missiles with chemical warheads. Unfamiliar with the consequences of chemical warfare, the nurses participated in a series of training lectures and were presented with video films of gas casualties. They were then assigned emergency posts at the hospitals and were instructed to report there as soon as they could on hearing the warning sirens. The war period, in contrast, was dominated by a soldier-like discourse of commitment. In the last period, after the war was over, that discourse was replaced with a rhetoric of betrayal and frustration. The following amalgamated narrative summarizes the changing circumstances and language of a single respondent, Miri, a head nurse in the nephrological ward.

First, [before the war,] the uncertainty was bewildering. Military experts told us about biochemical warfare, political commentators told us how mad and unexpected Saddam Hussein is. He [Saddam] actually voiced his threats against Israel on our TV. There were biochemical emergency drills at the hospital, and we were told that we were living in a "high-risk zone." Until then, I thought Tel Aviv was in the rear. Missiles, that's all we heard of. He's got missiles that can fly straight from Baghdad to Tel Aviv.

Then, during the war itself, it was different. We had to leave our families behind and go to the hospital whenever a missile attack occurred. It was a different feeling. We were doing something important, like soldiers. Our husbands were not called [up]. It was only the missiles and us, and we found ourselves defending the rest of [the citizens]. The hospital was a great place to be in, it really gave you a feeling of pride and responsibility, unlike the home, where everybody just listened to the TV and the radio. Then, the Iraqi missiles and the sirens became routine. Those missiles usually missed their targets. And they had no chemical heads, as they were called.

After the coalition forces won the war, again it was different. People began to say that it wasn't really a war. Were there mass casualties? Was there an invasion? It was like waking up from a bad dream that did not materialize. People were quick to forget all about it, and we felt like we were forgotten, too.

Miri's experiences and the changing rhetoric she adopted to understand them were common among the interviewed nurses, as the following analysis demonstrates.

Stage 1: Before the War

As the coalition forces gathered around the Persian Gulf and Sadam Hussein's threats to launch a missile attack on Israel were publicized in the media, Israeli society began a chaotic process of preparation. In anticipation of a gas attack, the government selected several large hospitals to handle what promised to be mass casualties. Showers were constructed in their parking lots, specific rooms and wards were designated to receive the gas victims, and nurses as well as physicians were assigned war duties.

Miri described her activities and her hospital's in our initial interview: "At the end of each working day, I walk near the showers built for the gas attack and recall my mother's stories of the holocaust—how they were all stripped by the Germans and washed in the showers. It makes me shudder. It was decided that my ward, which is located in the cellar, would be turned into a large emergency room. Life support systems were put nearby, oxygen masks, large quantities of atrophine, all that stuff.[5] I feel like a second holocaust is approaching."

These comments were echoed by other nurses. Malka, working in the newborns' ward, told me that her greatest fear was "from the chemical danger. I was practically not afraid of death, but if I was to be injured by the chemicals, if something would get complicated, how would I look afterwards, who would take care of the children? This was more frightening than death."

Panic in times of uncertainty is commonly nourished by "horror stories," subterranean rumors that are presented as true and are widely believed but are groundless (see Goode & Ben-Yehuda 1994: 108–12; Best 1989). In the case of the nurses, these rumors included stories about unavoidable and disastrous chemical warfare, Sadam Hussein's madness, and the unprepared Israeli population. The nurses frequently used the holocaust example to articulate their fears. But that example also served to propel them into action. Whereas the holocaust image continued to be raised by men during and after the war, nurses used it only in the first stage—before the war. It was, after all, their responsibility to make sure that a new holocaust would not take place.

Nira, a gynecological nurse, told me: "What scared me the most was the speech of the hospital's head. I recall one particular sentence of his: 'You will encounter thousands of casualties, but first take care of those who have the highest chance to survive. Those who do not stand a chance must be given less priority, even if they are children.' The idea that I would have to take part in a screening—in deciding who has 'high chances' and who does not—gives me many nightmares." Particularly telling are these remarks by Gila, a midwife nurse: "Chemical injuries are different from regular ones. We have no experience in chemical injuries. Nevertheless it is—as with any other soldier—my national, professional, human, and social duty to come to the hospital and take care of the people. I am a soldier." This final sentence combines both the fears of the first stage and the rhetoric of commitment that was to replace them during war itself.

Stage 2: During the War

"During the war we were the state's soldiers." This was the nurses' most common phrase when asked to sum up their war experiences. In practice, what did the nurses do during a war that had almost no casualties? Upon hearing the sirens, they had to put on their gas masks and walk or drive to the hospital, leaving behind, in many cases, their "sealed" husbands and children. Once arrived, they found that work was fortunately lacking. In fact, most of the casualties they treated were victims of "false injections" (as nurses referred to them), people who panicked and injected themselves with too much atrophine.

The fears of the period before the war gradually faded. Many nurses now referred to them as "something disseminated by the media." Instead, they now spoke of their sense of pride, commitment, and duty. As Dina put it: "I am like everybody else. 12 hours. From 7:00 P.M. to 7:00 A.M. Like a soldier. [What about the children?] The big one watches over the youngsters. My husband helps. [How does it feel to be 12 hours in the ward?] It gives me confidence. The hospital is like a bunker. Being with everyone all the time—. We make gorgeous meals, play cards, dominoes. We even sleep together. There is a very strong feeling of commitment, of patriotism. We do it for the land of Israel." Another nurse, Pola, said: "I feel like the hospital is the place for me to be during the war. This is my front. If I did not come because the kids at home need me, it would be like deserting my post. I must help those who need me. I am a soldier nurse."

Several nurses told me how they brought their children with them to the hospital. The result was often improvised kindergartens of up to 120 children. When I asked why they brought their children along, nurses emphasized what they perceived as the relative security of the hospital compared with the sealed room at home. Many of them said that they would rather come to the hospital during SCUD attacks than stay at home. At home, they said, the ambiance of war was stronger. This made sense, because during SCUD attacks, the home often became a chaotic information center, with radio and television loudly playing, and relatives and friends phoning. As Rina explained: "At home I was much more pressed; here at least I felt use-

ful. At home I was like paralyzed, looking out through the window and see-
ing no one. One time, the taxi driver who took me to the hospital in the
middle of the night told me that I was crazy to go out in a time like this. I
told him: 'I am a nurse.'" Similarly from Sara: "I was proud to be a nurse
and to have to come to the hospital. My husband stayed at home with my
mother and our children—and it was the first time I did not worry about
them."

The hospital was dually constituted during the war as both a second
home and a front. It was at the hospital that nurses "felt more secure," and
it was also at the hospital that nurses (to use their own terms) "fought" and
"protected the homeland." The two conceptions of the hospital served a
dual purpose. Regarding the hospital as "a front" cultivated the patriotic
feelings of nurses and helped them make their case to their husbands and
the rest of Israeli society. But the hospital was also a "second home," a place
where one spent 12 hours a day, cooked meals, played games, and talked
with friends. The hospital was also arguably safer. Hagar said she thought of
the nurses as some sort of a "rescue team in the middle of danger, a . . .
Rambo to whom nothing can happen." "Upon hearing the sirens we would
go to the battle, each where posted," said another nurse.

Here is a typical interview with a "soldier nurse." Me: "When you had to
leave home during darkness or a missile alert, were you not scared? Didn't
you think about yourself? About taking such risks?" Shoshi (age forty-five,
head nurse, married with two children): "No, I did not think about it. I
thought about other Israeli soldiers, like Trumpeldor, who said it is good to
die for our land." Me: "During missile attacks, whom did you feel committed to? First to the patients? The family? Yourself?" Shoshi: "I felt that my
first duty was to the land of Israel." Me: "How did your husband react to the
fact that you were going to work in a time like that?" Shoshi: "He did not
complain. It was clear that this was part of my duties. At times he got angry,
but I did not care. I explained to him that I deserve his support now for
going to the army reserves [milu'im] in his place." Me: "And the daughters?"
(Shoshi had two girls, ages eight and fourteen.) Shoshi: "They wanted me to
stay, but acted perfectly when I left."

Other nurses spoke about a feeling of "being chosen," a feeling of
uniqueness and common destiny. Carrying out commands was another
part of the soldiering discourse: "All I did was follow orders" was a common
phrase. Several nurses pointed out to me that they were warned that a pun-

ishment of up to three months in jail awaited "deserters" (a term taken directly from the military's jargon) who did not show up for work. This threat (like Sadam Hussein's) later proved to be false, a fact that elicited anger and disbelief from many nurses.

Many of the nurses expressed the same feeling of solidarity found in combat units around the world (see Higonnet et al. 1987), including the IDF's (on Israeli fighters, see, for example, Gal 1986; Katriel 1991; and S. Cohen 1997)—a sense of togetherness in a "sisterhood of fighters." If wars are rites of passage, then these descriptions of the second stage of the war could be compared to the intermediate phase of transition rites (see Van Gennep 1960), where those undergoing the rite experience a state of "communitas," to use Victor Turner's (1977) term. Many nurses said that they felt like the status differences separating them from the (mostly male) doctors became insignificant. Mina, a head nurse, said it was "wonderful." "People with whom you worked for 20 years and did not develop more than a superficial personal contact suddenly lost their distance and became closely attached. Like one big family. I came to know people inside and out during the war and this itself was a reason for celebration."

Not only gender and professional but also national boundaries were dissolved. Talila, an Arab nurse, told me how during the Gulf War, when she left her home in an Arab village in the middle of the night and headed for the hospital, cars of the Shabac (the Israeli internal security services) that were patrolling nearby would aim their lights so that she could better observe the darkened road. Before the Gulf War, Talila said, those cars would "shine their lights into my eyes in order to blind me."

When asked about their real home, nurses denied any conflicts during the war. After the war, however, other perspectives surfaced. Chani, working in the children's ward, reported: "Whenever my husband was at home with the kids, there was no problem concerning my work. The problem arose when he had to be out for his own work. Then I was in conflict, torn between my professional and national commitment and my duties at home. As a nurse I was committed to come to the hospital where people needed my help, but as a mother my child needed me just as badly. Finally, I had to leave my kid with the neighbors."

The home/work conflict is, of course, a gendered one. Israeli male fighters, it seems, have rarely felt torn in that way. The men Edna Lomsky-Feder (1994, 1998) interviewed, in her study of the war language of Israeli veterans, seldom referred to such a conflict. In their (and Israeli society's) eyes, their absence from home was a necessary and negligible byproduct of their military career "in the service of the nation." However, when the nurses attempted to appropriate this script, their voice was silenced. Israeli society did not recognize them as the soldiers of the Gulf War, and their story was papered over by the national, male-dominated text that denied the very existence of the war.

The home/work conflict (expanded to home/front in the case of nurses) always came up in the interviews. As one nurse put it, when she was "at the front," her children were undergoing "army basic training" (*tironut*), meaning that their staying behind at home was a sort of a strengthening rite of passage for them (notice, again, the appropriation of military jargon). Several nurses expressed a sense of shame for being more secure at the hospital than their family was at home. Most important, many of them told me that their husbands did not "really approve" of their absences from home. This disapproval became particularly evident when it gradually transpired that no casualties were arriving at the hospitals. The rhetoric of commitment and soldiering, therefore, was needed as a justification for going to work and leaving the children at home, under the responsibility of the father. The discourse of soldiering was no doubt chosen because it was both called for by the circumstances and already constituted part of the normative order of Israeli society. In addition, the commonly heard phrase during the war "at last our value [as nurses] is recognized" should be read against the historical background of the low-prestige, gender-specific profession of nursing, a topic I will return to in due course.

Stage 3: After the War

The end of the Gulf War provided a brief catharsis, followed by a sense of frustration. This was true both in Israel and worldwide.[6] The war had ended, but Saddam Hussein quickly reestablished control over rebel areas and probably resumed nuclear production. Radical critics of the Bush administration resumed their criticism.[7] In mid-January 1992, Saddam celebrated the anniversary of his "victorious war." Contrary to Jean Baudrillard's (1991) assertion that the Gulf War did not take place, there *was* a war, and it was a very real one as far as the Israeli public was concerned. It had a profound impact on Israeli life, shattering the faith in the military, the political leadership, and the illusion of national independence (Israel was pressured into deferring completely to American interests and prescriptions during the war).

The frustration felt by nurses after the war should therefore be considered in the general context of the war's aftermath. The following quotations will serve to illustrate this sense of frustration. Gila, the gynecological nurse we met earlier: "How do I feel after the war? I feel empty. As if I was completely emptied, and then left alone. A feeling of exploitation. I refer to the system that demanded that I work during the war, while completely ignoring the fact that I was a woman and a mother, recently married and with children." Orli, an emergency room nurse: "It was a tough and difficult time. I don't like to recall it. I don't like to discuss it. What I want is simply to close down this chapter of my life, and forget it ever happened." Naomi, assistant head nurse: "It was chaos. My faith in a strong and omnipotent Israel was destroyed."

The last remark could have been heard from any other Israeli citizen after the war, but the other two disclose feelings peculiar to the nurses. In particular, they reflect their frustration over the fact that, in the words of several, "no one appreciated our tremendous efforts." Most of them ascribed this lack of appreciation to the general dismissal of the nursing profession in Israeli society. A few associated it with the general frustration felt by the Israeli public after the war. As one put it, "In a situation where everyone's frustrated, no one pays attention to the other's sacrifices." Many nurses complained that as things returned to normal, no one—their husbands, the male doctors, Israeli society in general—seemed to remember what they had been through. They were especially angry with the doctors, who "immediately resumed their arrogant ways and looked at us from above."[8]

Nursing, Soldiering, and Subversion

The tradition of "men in arms and women at home" is one of the most prevalent in human cultures and histories. It is also a fiction, one that is usually told by men. Some scholars have used the experience of the Israeli nurses in the Gulf War to shed light on some of the premises and consequences of this fiction. Miriam Cooke, for example, claims that the Gulf War was a "postmodern war":

Whereas wars previously codified the binary structure of the world by designating gender-specific tasks and gender-specific areas where these tasks might be executed, today's [postmodern] wars are represented as doing the opposite. Postmodern wars highlight and then parody those very binaries—war/peace, good/evil, front/home, combatant/noncombatant, friend/foe, victory/defeat, patriotism/pacifism—which war had originally inspired. (1993: 182)

Cooke's general assertion, which was argued in a Western-feminist context, needs to be further located in the Israeli context, particularly in regard to the local meaning of the cultural script of soldiering. Israel is used to having recurrent wars that interrupt its social order. This abnormal-cum-routine phenomenon was termed "the interrupted system" by the Israeli sociologist Baruch Kimmerling (1985). Within the constantly interrupted Israeli system, war and peace are (or at least have been) constructed not as a matter of either-or, but rather as a complementary reality. This has no doubt had immense sociopsychological repercussions in terms of justifying militarization (Ben-Eliezer 1995) and propagating a "siege mentality" or "Masada complex" (Nachman Ben-Yehuda 1995).

As a result, military service is a "key symbol" (Ortner 1973) in Israeli society, denoting manhood and legitimizing male superiority. Amiya Lieblich (1979: 15), who studied the ethos of Israeli soldiers, found them typically saying: "War is an integral part of our life in the country. At war, men in arms risk their lives in order to defend the citizens. Women, the elderly, and children must stay behind and worry about the fate of their relatives. This is how it must be."

The militaristic script arguably provides the larger context for gender construction in Israel. As noted earlier, military service, traditionally a male domain, still largely defines the extent to which an individual is in the "social-evaluative" system of Israel (Horowitz & Kimmerling 1974; Gal 1986; Ben-Ari 1989). Arabs are not drafted, and until recently, women could not serve in combat units.

Wars (postmodern or otherwise) often create a state of emergency and "communitas," thus enabling role (including gender role) reversals. The example of the Gulf War nurses is one illustration. But Israel has a long tradition of female fighters, from the palmach (prestate elite troops) to present-day IDF tank instructors. Nevertheless, these women's roles were (and still are) generally bracketed as exceptional and symbolic (Yuval-Davis 1985). While the Cooke quote illustrates how fluid and resurgent definitions are, the militaristic script of Israeli society explains why the residues of gender remain largely unchanged.

Furthermore, the term role reversal is a biased and incomplete description of what happened to the Gulf War nurses. To merely describe the tasks these wives and mothers took on as a role reversal impelled by the war would be to subscribe to the male claim that women can be either (normally) at home or (abnormally) at the front. What actually happened, as we have seen, was that nurses did a "double shift," at home and at work—as is often the case with women, particularly working mothers (Hochschild 1989). Given the propensity of men to take women's "double shifts" for granted, as well as the construction of soldiering as a male domain in Israel, it is not surprising that the Gulf War nurses were largely forgotten by Israeli society.[9] Add to that the fact that for many men, their nonparticipation in that war was totally galling. Consider, for example, this observation:

> Israel's passivity in the Gulf War was especially frustrating to the male reservists who, instead of being called for active military duties, found themselves waiting in sealed rooms together with their families, having little to do except for complaining about their ill-fate as unused fighters. This sense of humiliation was probably reinforced by the observation that the main active duties during those times were typically performed by the women rather than the men of the family. Thus, for the first time in their war-laden history, Israel's incapacitated men were watching their wives undertaking most of the responsibilities for protecting the safety and welfare of the family. (Yuchtman-Yaar et al. 1994: 7)

The overlooking of the war contribution of nurses has parallel examples elsewhere, the most notable being the indifference to the role of American nurses in Vietnam. Many nurses (estimates range from 4,000 to 15,000, according to Norman 1990: 4), most of them serving in the Army Nurse Corps, were stationed between 1965 and 1973 in over 30 hospitals and intensive care units in South Vietnam. No official record of this group as a separate unit was kept; "no one thought that researchers would be interested in this group" (ibid.). The U.S. Senate Committee on Veterans' Affairs, for example, did not see fit to include those women in its broad-ranging research on Vietnam veterans (ibid., p. 134). Indeed, until the mid-1980's, even the nursing profession itself failed to acknowledge the Vietnam veterans in its ranks. Moreover, to many of those nurses, the war provided a time of responsibility and authority, and it was difficult for them to surrender afterward to the male-dominated medical system of the civilian world. The overall similarity with the Israeli case highlights the issue of gender. In peacetime, each gender typically performs specific roles; in wartime, nurses are able to take on roles previously reserved for men. (For more on the Vietnam nurses, see Freedman & Rhoads 1987.)

The appropriation of the masculine discourse of soldiering by Israeli nurses during the Gulf War should therefore be seen as an adoption of a culturally available rhetoric that could make sense of their changed circumstances and experiences. Once having adopted this rhetoric, some of the nurses no doubt recognized it as a means for garnering prestige in a profession traditionally seen as embodying "passivity, self-sacrifice, devotion, and subordination, . . . all unambiguously [located] within patriarchally constructed femininity" (Gamarnikow 1991: 110). "The physician or surgeon prescribes, the nurse carries out": this phrase, written by the founding mother of modern nursing, Florence Nightingale (1882: 6), is still a binding one. "Faithful servants we should be, happy in our dependence, which helps us to accomplish great deeds" (former matron, 1906, cited in Gamarnikow 1991: 118). Such a discourse of obedience and service is not very different, in principle, from the discourse of soldiering. The characteristic features of nursing (low pay, low prestige, bad hours, high turnover, and lack of job autonomy) are determined by a single feature of its practitioners, namely, their gender (see also Ehrenreich & English 1976; and Game & Pringle 1983).

In interviews, the nurses clearly spoke about the low prestige of their profession and the need to rehabilitate the nurse's image. This view was a widely shared one and arguably reflects a group consciousness derived from a similar educational background and work conditions. (Academic education for nurses was introduced in Israel in 1968, but until 1982, all that was offered was a post-basic course for registered nurses. In that year, Tel Aviv University instituted a master's program; Shuval 1992.)

Despite the nurses' frustration over the low prestige and low salaries of their profession, the group consciousness they displayed can also give rise to idealistic notions—such as "soldiering." But it takes real determination. Past attempts by Israeli nurses to improve their working conditions and salaries were regarded as immoral efforts "to raise the price of an essential service and turn what ought to be a mission into a business" (Ben-David 1972, cited in Bloom 1982: 158). As Israeli society sees it, Joseph Ben-David (1972: 36) concludes, "the only way the supply of manpower can be maintained, given such a set of circumstances, is by using cheap female labor and appealing to certain kinds of idealism to which women are sensitive." This is true not only in Israel. In the past, for example, American recruitment posters for wartime nurses always emphasized the saintly, spiritual image of the nurse (sometimes showing her in the same pose as the famed Pieta; see Kalisch & Scobey 1983: 221). Never in American history have legislators raised the issue of exposing nurses—even civilian ones—to the high risks of combat areas; the implicit assumption is that women, "by virtue of being nurses and because of the exigencies of war, [are] exempt from prevailing protective attitudes about the utilization of women in combat areas" (ibid., p. 238).

Because the rhetoric of soldiering is part of the Israeli culture of the chosen body, a national-militaristic discourse that serves to promote the enlisting of the citizen for national goals, a complementary reading could regard the nurses' adoption of that rhetoric as a pragmatic means they employed to justify—in their own eyes as well as their husbands'—their absence from home during the war. After the war, as national pride faded away, and there was no time for any discourses of soldiering, that rhetoric was no longer useful. Nurses began to question the very legitimacy of the country's "sending them away from home." When denied the symbolic capital related to the manly script of war that their rhetoric appropriated, many nurses turned their postwar frustration into a critical reexamination of the key-symbols and key-scripts of Israeli society. This disillusionment with the fundamental postulates of collective life is perhaps characteristic of periods following unvictorious wars, as was the case in the United States after the Korean and Vietnam wars, and in Israel after the 1973 October War (Livne 1977).

In sum, nurses adopted the masculine rhetoric of soldiering because of the availability and instrumentality of that discourse, as well as the absence of another set of symbols. "Soldiering," judging from what seemed to be an authentic enthusiasm during the war, was something the nurses clearly felt with all their hearts. It would have been only natural had they chosen the image of the "national mother," derived from the motherly ethos of nursing and caring, as their symbol of commitment during the war. No doubt they could have followed Sara Ruddick's (1989: 132) dictum of "maternal thinking" and generalized "the potentiality made available to activity of women, i.e., caring labor—to society as a whole." But they adopted a different image instead—the militaristic image of the soldier-nurse. (Interestingly, during the war, many nurses used a masculine plural form in referring to their activities: *anachnu mesachakim*, we are playing; *yeshenim*, sleeping; *ochlim*, eating; and so forth. But after the war, they reverted to their usual feminine speech.)

All in all, the situation of the Gulf War nurses was rather similar to that of the Israeli female soldier in general. Both receive a double message—they are involved in war, on the one hand, but they are denied the "privilege" of being "real soldiers" on the other. Women's lives, as nurses or soldiers, are socially circumscribed to allow them military service, after which they must return to society understanding their role as a woman. This argument fits Dafna Izraeli's (1997) elaborate model of the IDF as a gender regime (following Connell 1987, 1995). Izraeli argues that the military, as a gendered division of labor, constitutes and maintains the taken-for-granted role of women as helpmates to men. As such, the military intensifies gender distinctions and then "uses them as justifications both for their construction in the first place and for sustaining gender inequality" (Izraeli 1997: 129; see also Levy-Schreiber & Ben-Ari 1998).

A. R. Bloom (1982: 157) sees this situation as stemming from the Jewish tradition, going back as long as biblical times: "It's good for a Jewish woman to be strong and aggressive when the Jews are in danger and she's acting in the people's interest. If we go through the Bible and legends carefully we see that whenever Jewish survival is at stake, Jewish women are called upon to be strong and aggressive. When the crisis is over, it's back to patriarchy." The militaristic script, although usually regarded as differentiating Israel from the diaspora, the new sabra from the old Jew, is therefore also part of the larger Jewish tradition, which has always separated the roles of men and women, designating the public domain for the former and the domestic for the latter (Heilman 1983; Sered 1990, 1992). War has long been a male space, and its stories have therefore been told, and its meaning decided, by men alone. This chapter attempted to retell the story of the chosen body in a different, gendered voice.

In January 2000, the Israeli parliament passed a law permitting women to enter combat units. This law, however, is not going to turn the military upside down, since recruitment to elite combat units still depends on the prescreening of male candidates alone. Furthermore, actual recruitment patterns are subject to military policy and are determined by ranking officers. To be sure, the new law has allowed women to compete for combat positions, and there are now some women pilots, missile ship captains, and artillery operators, for example. But they constitute a very slim minority, less than 2 percent of the total in all three occupations. The chosen body, therefore, is not at a crossroads. The military will continue to be dominated by male (chosen) bodies in jungle green. Furthermore, the (very slow) entrance of women into the world of the chosen body continues to be under conditions set by men, at the convenience of men, and under the judgment of men.

5 THE CHOSEN BODY AND THE MEDIA

LIKE ALL DISCOURSES of normative control, the culture of the chosen body relies on its agents of reproduction. These include the family (discussed in the context of screening fetuses and newborns and socializing youth, as well as in the specific context of bereavement) and the medical system (discussed in the general context of testing and screening fetuses, babies, and soldiers, as well as in the specific context of the Institute of Forensic Medicine and state-run ministries and agencies). This chapter discusses the role of the Israeli media in reproducing the culture of the chosen body.

The mass media have a significant role in shaping collective representations of and reactions to such public issues as terrorism and militarism—two issues that not only define the boundaries of the Israeli collectivity, differentiating between "us" and "them," but are at the very heart of the culture of the chosen body. The criticisms of the American media's approach to terrorist violence, for example, have been described as emanating from their symbiotic relationship with the government. The media help to mold foreign policy by parroting government constructions of terrorist actions, and thus influence public opinion.[1]

Behind the "domination of the state through fear of terror" thesis, one can find the seminal study of the media by Edward Herman and Noam Chomsky (1988; see also their ensuing critique in Chomsky 1989; and Herman 1992). These authors analyze what they see as a systematic media bias in the United States and provide a propaganda model to explain its occurrence. They contend that five filters act to screen out news, marginalize dissent, and give dominant public and private interests undue access to the public: (1) concentrated ownership and the profit orientation of the mass media; (2) media's dependence on advertising revenue; (3) media's dependence on government, corporations, and elite experts for information; (4) media's responsiveness to negative public reactions; and (5) media's anticommunist orientation. All these filters were quite relevant to the Israeli media until the 1960's, when things began to change.

The Mass Media in Contemporary Israel

In the 1950's and 1960's, Israel's one and only TV channel and big daily newspapers were in the "manufacturing consent" (to use Herman and Chomsky's term) business. For example, they contributed mightily to the feeling that the "cold war" with the Arabs was permanent, and that terrorist attacks must be emphasized to maintain national solidarity (Nossek 1994). But as Tamar Liebes (1998, 1997) points out, there has been a drastic change in recent years, "with the monopoly of public broadcasting giving way to a multiplicity of fiercely competing channels, and the technological revolution facilitating live transmission from a multiplicity of points" (Liebes 1998: 72). In the 1990's, Israeli leaders faced increasingly harsh vocal opposition as the parties responsible for the renewed terrorist attacks by those who were bent on thwarting the peace process. Following a series of 1996 bombings, the Labor government seemed "incapacitated, its credibility badly damaged. . . . The right-wing opposition who, until the attacks, had lagged behind in the polls framed the event as the outcome of the government's peace process" (Liebes, n.d.: 14; see also Liebes & Peri 1997). In the end, the Labor government lost the May 1996 elections and was replaced by the right-wing Likud party, headed by Benjamin Netanyahu.

At this stage, the media ceased to sit passively by as agents of consent manufacturing. On the contrary, they became more and more active in bringing to the surface fundamental problems that promised to undermine the political system. Liebes (n.d.: 9–10) emphasizes three major issues that were discussed on the screen and in the print press following several disruptive incidents in the 1990's. One issue was the previously unquestioned reliance on the resourcefulness of the military. Another was the division between the majority of Israeli youth who do three years of compulsory military service and the orthodox youth who have full rights as citizens but are not conscripted. The third issue was the continued presence of the IDF in southern Lebanon.

The media, in short, assumed an active part in the political debate, airing different voices and issues. At the same time, the media also kept to their more "national," unifying role by following the three-phase script of loss, regathering, and (ultimately) recovery. But far from unifying the country, their coverage in the very first phase, of loss and chaos, brought them under heavy criticism for their nonstop "disaster marathons" (as Liebes calls them) of terrorists attacks. The two TV channels were actually accused of "inciting hysteria, of losing all proportion in repeating the attacks, of competing over the degree of aggressiveness" (Liebes 1998: 71). One political scientist went so far as to charge TV's senior news anchor with "acting as an agent of the Hamas" (ibid.).

In what follows I treat the media coverage of three events: the helicopter crash in 1997, a series of terrorist bombings in 1996–97, and the assassination of Yitzhak Rabin in 1995. My emphasis in dealing with these events is bodyTalk—how they were described, explained, and given meaning through what is primarily a rhetoric of the chosen body.[2]

The Helicopter Crash

On the evening of February 3, 1997, two army helicopters set out to transport troops to Lebanon on a routine exchange of personnel in Israel's "security zone." For maximum security, the helicopters flew close together and without their lights on. Approaching Israel's northern border, they collided and crashed to the ground, killing everyone on board. It was later reported that many of the explosives the troops wore on their bodies went off in the crash. Seventy-three soldiers were killed. In a commemorative story a year later, the daily newspaper *Yediot Acharonot* (30.1.98) returned to the scene of that national trauma:

> In the places where the soldiers fell, . . . marks have been left on the ground, which was on that night wet and soggy. The impressions of

three bodies were extremely apparent in Gershuny's [one of the inhabitants of the village where the helicopters went down] grass-covered yard. After three months, Gershuny took a shovel and covered the marks with soil, because he could not live with the sight any more. But the pilgrimage still continues. Every school excursion stops at Gershuny's yard.

This commemorative piece concentrated on the remaining marks of the bodies as an aptly embodied metaphor for the lingering, traumatic recollection. Moreover, it was not only the people who remembered, but the land itself. The land marks had become "landmarks," "body marks"; they were the real war monument, "places of memory" made sacred through grass-roots pilgrimages. I have already touched on the crash in Chapter Three to portray the collective discourse of bereavement in Israeli society. Here I return to it to show how the media coverage highlights the use of bodyTalk that surrounds contemporary issues of terror, militarism, and sacrifice.

Media Obituaries

As in the case of other recent national disasters, the two nationwide TV channels and the nation's radio channels, both public and commercial, interrupted their regular schedules, canceled all advertising, and devoted all their time to the event for three full days. The broadcasters typically first reconstructed the event, reported that there were no survivors, and announced the decision to set up an investigating committee. They, and the newspapers, then turned to the mourning, providing the life stories for the long list of names. This mood, sad but somber, replaced the usual "chaos" stage that follows such national traumas in Israel. I describe this stage through one of its primary constituents—media obituaries.

"Media obituaries" are, unfortunately, a familiar genre in Israel. They are short texts, usually with a picture, that are run in all the newspapers and are written by their staff reporters. The obituaries following the helicopter crash were especially prominent; because of the enormous number of losses, they filled the whole front page of each daily Hebrew newspaper that day. In one case, the front page later became a metonym of the trauma; the following day, the newspaper published on its front page a picture of soldiers in active service looking at the last day's front page, obviously trying to identify familiar faces.

The crash victims' obituaries were typical of their genre, and their analysis will therefore provide a window into the bodily characteristics of that genre. Overall, it is a genre through which the chosen body can be reconstituted and endowed with symbolic immortality. The fallen soldiers are always portrayed, first and foremost, as a chosen body—handsome, brave, fit, well trained. When journalists write about the fallen as a group, they usually refer to them as "our beautiful boys." When relatives are quoted, the reference is more personal: "your beauty was gorgeous," "you had the most beautiful eyes ." Friends say that "he was a real fighter," "a real man." Another common phrase is *with his body he protected our borders* [italics mine]." The chosen body of the fallen stands for the body social and the contours/borders of the territory. Body and territory become one.

The adjectives attached to the fallen usually portray their special quality: the fallen was "exceptional," "one of a kind," "perfect," "a perfectionist," "talented," "the best." These qualities are "proved" or given extra emphasis through a common narrative describing the difficult path to combat service and the deceased's determination to succeed in it. One crash victim's obituary claimed that "he came to Golani [an infantry division] with an ulcer, but did not tell anyone, and finished the course with distinction." In a similar vein, a family member noted, "he was hit twice in Lebanon, but didn't give up. Didn't tell us, kept it a secret, so as not to be taken away from his combat unit." Another soldier "was conscripted last year with a low medical profile because of asthma problems. Immediately he began his own war to raise his profile. He wanted to serve in the most risky combat unit there is." A typical preliminary part of this postmortem chosen-body narrative is a successful high-school career in sports: "He was a distinguished athlete" is a very common description.

Because the crash was the result of human error (caused by "our own hands"), it also required self-criticism and contemplation. Even these "negative" comments, however, served to highlight the chosen body as the ultimate and unshaken reference point framing the whole discussion. The body of one of the helicopter pilots was autopsied at the Institute of Forensic Medicine.[3] On the face of it, the pilot was a "chosen soldier" belonging to one of "the finest" (the Israeli air force). But he was also allegedly at least partly responsible for the horrible crash. That is probably why the media gave wide attention to the results. The doctors found, according to one account, that "some of the pilot's arteries were in bad shape. . . . Medical sources were surprised that he was permitted to fly with such a high level of cholesterol in his blood."

The media's bodyTalk also concentrated on the tearing of soldiers' bodies and the exploding of the helicopters. Pictures focused on body parts of soldiers and on ruptured metal parts from the helicopters, as well as on torn bits of bodily equipment (backpacks, uniforms, military shoes, etc.). The metal fragments pictured were described as "having screaming faces." The helicopter's remains were likened to body organs, with particular emphasis on "the tail" and "the stomach" (where the soldiers were carried, "like babies in a womb").

BodyTalk was also prevalent in the comments of officers describing their feelings at the funerals. A reserve officer told a journalist that "10 years in the job did not immunize him," and that whenever he had to break the bad news to the family, "my guts begin to roll and my throat becomes immediately dry." Another said, "Each time before I knock on the door [of a bereaved family], I get goose bumps. My heartbeat accelerates, and I sweat." "The crash opened up wounds that seemed to be healing so well," Amram Mitzna, a former high-ranking commander and mayor of Haifa, told a newspaper reporter. He recalled how his "hands trembled. It seemed that these hands that touched so much death had a life of their own. The hands that touched bereaved mothers, that operated tanks and artillery, these hands could not stay still. My hands trembled as I awaited a telephone call from my youngest son who serves as a parachute officer."

Finally, we find the media applying bodyTalk to nature itself. The land is personified, given a body, as it remembers the fallen. The land "has eaten" or "swallowed" the soldiers. The crash has caused an "earthquake." Nature, and more specifically the national territory (Israel is "weeping" together with its "sons"), thus echoes the tragedy. "The terrible explosion that caused the crash has lit the skies with a great light, but our world is darkened." And the newspapers even report, "in the morning, the sun rose and shone, but did not give warmth." In another obituary, the reporter wrote, "the cold and bitter night will not be forgotten . . . here or in the other bleeding parts of this land."

Regathering and Recovering the Chosen Body

Here again, in the second stage, the media coverage of the crash was typical. In a parallel manner to the rupturing of the body—personal, social, and territorial—this stage consists of the regathering of those three bodies. First, the personal body was regathered: the media documented the gathering of body parts and remains. The pictures of horror and chaos were now replaced by pictures of closure: the sealed and standard wooden coffins used by the military. The media no longer portrayed body parts, and when individuals were mentioned, the accompanying picture was a portrait before the accident took place. The bodies, identified, collected, and sealed inside coffins, are put to rest.

Next (or simultaneously) came the regathering of the social body. Politicians and journalists called for a "return to normal life." As a graphic illustration, the media showed pictures of the reassembled helicopters. We were informed that air force experts were able to put "the wrecked pieces side by side in the original order," so that they could be "inspected by the military experts and by the Committee of Investigation." Inspection and rational learning had thus come to replace chaos and uncertainty. We see the "wounded" helicopter after it was fixed; we even see it in the air again. One picture showed a soldier affixing a new propeller. These pictures stood in stark symbolic contrast to the pictures of the helicopter parts ("the tail" and "the belly") seen only a week before. There followed reports that the "relevant military units are already back to normal activity," and the "helicopters are back flying soldiers to Lebanon." Another symbol of the social body, the bereaved families, were also portrayed as "returning to normal activity." The media gathered all 73 families for a "one family" picture. There were reports on the symbolic "adoption" of the bereaved families by other bodies. For example, the "70 families who live in She'ar-Yishuv [the settlement where the accident took place], joined by three other families, have decided that each will adopt one bereaved family of the 73 families who lost their sons in the terrible accident." The "wounded" military units that lost many men were also "reassembled" as new recruits stepped in to fill the gap.

Finally, the territory was recovered; nature, the "body of the land and of the people" (*guf ha'arretz ve'ha'uma*), was also returning to normal. The media conveyed this notion first and foremost by emphasizing the positive and strong linkage between the fallen and the country. That linkage, which also appeared in the obituaries, reappeared in a way that struck a chord of closure. The media reported, for example, on visits of the president (a symbol of the nation-state) to the bereaved families. Of particular relevance to this discussion were the president's remarks that "these fallen soldiers are the flesh and blood of this country; they represent the good old country" (*Eretz Yisrael ha'yeshana ve'hatova*). With these reports, the media ritual of loss, regathering, and recovery (the latter two practically combined) had once more been completed (until the next time).

Terrorist Bombings: The Linkage Between the Body Politic and the Individual Body

This section discusses the media coverage of five terrorist attacks that took place in 1996 and 1997, whose media coverage embodied a similar three-phase process of loss and chaos, regathering, and recovery. The terrorist attacks that took place afterward followed the same pattern, which is illustrated here. My analysis of the coverage shows that bodyTalk was, once again, prominent in the reportage of each stage. The dynamic of terror was therefore also captured, explained, and accounted for through the body.

All five incidents were planned and carried out by Palestinian members of Hamas, a Muslim fundamentalist faction opposed to the Oslo peace process (and indeed to the very existence of Israel). Before proceeding to the press coverage, let me give some of the particulars of each. On February 25, 1996, two suicide bombers struck in Jerusalem and Ashkelon, blowing up a bus and a soldier's hitchhiking post. Twenty-four Israelis, two Americans, and one Palestinian were killed; 80 people were wounded. On March 3, 1996, a suicide bomber blew up a Jerusalem bus, killing 18 and wounding 10. On March 4, 1996, 12 people were killed and more than 100 wounded by a suicide bomber outside Dizengoff Center in Tel Aviv. On March 21, 1997, a suicide bomber in Tel Aviv's Café Apropos killed three women and wounded 42 people. On July 31, 1997, a suicide bomber blew himself up in the Jerusalem marketplace; 13 people were killed, and many others wounded.

Loss and Chaos

In a story headlined "A Market Turned Upside Down," the *Jerusalem Post* (31.7.97, p. 2) described the site of the terrorist bombing in the city's *mahane yehuda* marketplace as chaotic: "the streets littered with overturned boxes of fruit and vegetables, splattered with blood. Wooden boards used to transport the wounded now lay thrown to the side of the road, covered with their blood. . . . The smell of death, charred bodies, and burnt blood filled the air."

The smell, a distinct bodily issue, appeared in many other descriptions. Another *Post* story in the same issue (p. 3) was actually titled "There Was a Smell You'll Never Forget." It was the stench of death and explosives that mingled with the smell of fresh baked goods on Jaffa Road, just minutes after the explosion.

Still trying to capture the horror of the moment, the paper (p. 2) abandoned its usually calm tone to spice up its coverage with vivid accounts of personal tragedy: "Levy Kadoorie, a retired bus driver, and his brother Avi pushed their way through the crowds and past the security personnel that were trying to establish a semblance of order. They discovered that their father, Yosef, had moved his banana cart from its usual place and was spared injury. But their mother, Simha, was missing. The brothers scanned the darkened faces of the dead and injured but did not see her." This story emphasizes the leitmotif of the first phase: chaos, loss, missing bodies, and unidentified strangers. As with the helicopter accident, the rupture of the body was transferred from the private body to the surrounding body (the marketplace) and then to the body social and the territory. The media coverage focused on ruptured bodies and on the total chaos in the marketplace. Furthermore, the physical chaos was described through bodyTalk; the marketplace was described as having a "body" of its own. The internal organs of the marketplace—stands, products, and so on—were mingled with the private bodies of people. "A horrible sight in which body parts and broken vegetable stands fly around . . . with heavy smoke hovering above everything." Or, "the sheds were burning, and the floor was filled with a mix of watermelons and blood." Excretions (smells and blood) of the people and the marketplace blend: "It was a horrible blend of a smell of blood, gun powder, spices, and pickles." One of the survivors described how he "felt that the whole street was shaking. . . . Stands were flying in the air like pieces of paper, burying people underneath. I heard terrible screams and saw parts of bodies and blood everywhere." Finally, the marks of the explosions were described as "*tattooed into the sidewalks of the marketplace*," in much the same way as the bodies of the victims of the helicopters were impressed into the ground (italics mine).

The Dizengoff bombing was especially ripe for bodily idioms of loss and disorder. As a mall of considerable attraction in the center of Tel Aviv, it was "the wounded heart" of the city. The poet Chaim Hefer published a commemorative poem in one of the daily newspapers titled "Dizengoff Is

Bleeding." "A Horror in the Heart of Tel Aviv" cried a headline. Dizengoff was called "the heart of the first Hebraic city, and hence the pounding heart of the whole country." Other newspapers spoke in terms of different body parts: "the belly button of Tel Aviv," "the ultimate rear." In a story colorfully titled "A Stake in the Heart of Tel Aviv," a *Jerusalem Post* reporter (5.3.96, p. 3), thought first in terms of a private body: "Tears streaked the festive make-up on the face of 16-year-old Ortal Hershkovitz as she rushed away from Dizengoff Center following the murderous explosion yesterday afternoon." The typical description of chaos and disorder ensued: "We all just started to cry and run away. We didn't want to look at the bodies. We came shopping with two other friends; we don't know where they are right now." "The sights and sounds at Dizengoff Center," the reporter wrapped up, "had an awful familiarity: the bodies strewn on the street covered with blankets, the screams of the ambulance sirens, the orthodox volunteers climbing ladders to fetch body parts."

The bombing of buses, like the crashed helicopters or the ruptured marketplace, provided a context for the personification of place. Reporters spoke of shattered buses, "burnt metal skeletons," "a torn ad that was once on the bus," and "ruptured tin pieces." Following the bombing of the No. 18 Jerusalem bus and the attack at Ashkelon, Israeli TV canceled all scheduled programs and launched into 72 straight hours of coverage. Liebes (1988: 77) describes this "disaster marathon," typical of the first stage, as "recycling blood, tears and vengeance: all during the time-out, daily commercials and promos were replaced with the recycling of bloody pictures and sounds, tightened to retain only the juiciest lines. Showing the same wounded child, another horrified witness, or a bereaved parent on the way to identify a daughter's body worked to arouse anxiety and, at the same time, to provide the visceral pleasures of all soft-core pornographic genres." The marathon was even followed by a special logo featuring the total chaos of the city square moments later: "covered bodies, broken bus parts, masses of police officers, ambulances, the fire brigade; a big blonde woman crying, looking for her son; a young man who saw people without arms, without legs, without faces; a young girl who saw the head of a woman on the pavement, her body on the street; a religious soldier of the coroners' corps collecting scattered pieces of skin and bone from the trees" (ibid., p. 78). At the assigned time of the major evening newscast, this special logo replaced the traditional one, to symbolize that "all order had been destroyed" (ibid.). Liebes' analysis of the disaster marathon on national TV is completely in line with my description of the loss and chaos that characterized the first stage in the newspapers.

The typical motif of this phase—disorder and unidentified bodies—was given an unusual twist in the first report of the Ashkelon bombing; the terrorist, whose identity was still unknown at the time of publication, was reportedly wearing an IDF uniform and managed to mingle with soldiers without drawing attention to himself. This report (*Jerusalem Post*, 24.2.96) had the impact of further disrupting the identifiable features of the "Israeli tissue." The terrorist had now become "one of us," an insidious virus, unidentified by the body politic's immune system. In symbolic terms, this represented a forbidden mix of "us" and "them." "The fact that no one noticed him suggests he was in disguise," Assaf Hefetz, the police chief of staff, told the reporter. As a symbolic gesture, Israel had immediately closed off the occupied territories. This was a ritual designed to literally distinguish between "us" and "them"—ritual, because as Hefetz said, "while sealing off the territories reduces the chances of attacks, it does not prevent them entirely. This is a free country, people move around freely and there is always a chance that a terrorist can cross the Green Line or the Erez checkpoint, either on foot or in a vehicle even if the territories are closed." Uncertainty and disorder therefore reigned again. And in the late afternoon hours, after all the carnage had been cleared away, volunteers from the Chevra Kadisha's identification unit were still scouring nearby fields for body parts and belongings. The crowd of demonstrators dispersed after around two hours. At 6:30, around 40 people arrived at the junction and quietly lit memorial candles. The next day, the *Jerusalem Post* (26.2.96, p. 6) published an opinion piece headlined "Hit at Our Very Heart." The opening sentence read: "One picture of yesterday's horrendous bus explosion in Jerusalem showed a paratrooper's beret lying in a pool of blood."

Regathering

The reopening of a subdued Dizengoff Center on March 6, 1996, was reported in the *Jerusalem Post* (7.3.96, p. 3): "two days after a suicide bomber tore open its heart, Tel Aviv made a conscious but hesitant effort to resume normality. Dizengoff Center, though still licking its wounds, opened its doors again." Though "the false heartiness of seasonal sales could not restore the broken spirit," the regathering process had begun. The representatives of order were in place, and this time it was not to curb chaos: "Policemen, some in khaki, others in blue, formed the biggest groups as they guarded the entrances and patrolled the sidewalk to restore public confidence. Workmen on ladders were busy repairing the twisted entrances and shattered glass walls of Bank Hapoalim and Bank Leumi. The gutted automated tellers were covered with soggy cardboard; a city dumpster was filled with broken glass and twisted light fixtures."

The second stage of the coverage was characterized by key terms such as closing, closure, evacuation, inspection, investigation, and identification. The city, as a body that was wounded in the bombing, gradually resumed its normal life, its "main arteries opening up for traffic." The newspapers' description of these reconstruction activities were part of the formula of loss-regathering. The manner of reporting is hence not just informative but ritualistic and symbolic. The marketplace, described after the bombing as a "torn, scarred body," was being "rehabilitated and renovated." The newspapers followed up on the shattering of Bus No. 18 with descriptions of the clean-up: "Everything was cleaned quickly, the bus skeleton was taken, the road was cleaned, purified." The special orthodox staff responsible for collecting body parts so the dead could be buried were working at the scene, "evacuating the pieces, regathering ruptured body parts. Slowly, they clean[ed] the place of any human remains" "In their usual meticulous work, the "true mercy' [*chesed shel emet*] staff collected from the floor torn limbs and pieces of skin that had flown out to more than a ten-meter radius. . . . They used ladders to climb up electric poles and scrape flesh off them." (The "mercy of truth" men, employees of Chevra Kadisha, with their plastic bags in hand, are a morbid symbol of the second stage.) The Institute of Forensic Medicine continued the work of regathering (called "reassociation" by staff members). Professor Hiss, head of the institute, reported in the newspaper: "We checked the bodies according to a certain order and a list of identifying features we collected. . . . I asked the families to arrive with photos and provide further identifying details. There was no difficulty in completing the process." When relatives insisted on verifying the corpse, however, social workers tended to prevent it, explaining that "the family should remember their dear ones . . . beautiful, whole, and nice." Corpses prepared for burial were wrapped in shrouds (contrary to the usual Jewish custom) so that the body could be remembered as intact. Sometimes bodies were wrapped in the national flag.

The day following each bombing was declared a national day of commemoration; this was part of the ritual of regathering. The group, the collectivity, also regathered. A "minute of silence in memory of those killed in the bloody #18 bus bombing opened today's school day around the country." According to one newspaper account, Dan Tichon, the Knesset's chair, comforted his secretary, who lost her son in the marketplace bombing, by telling her: "Now you are an inseparable part of our family; your son has already become our son, the chosen son of the Knesset." The president, Ezer Weitzman, was quoted as saying, "We are strong and should remain strong. . . . They must not be given the satisfaction of having wounded our body, the body of Israel. We must keep our backs straight and live our lives as usual." Famous writers joined in this public regathering of the "Israeli spirit." As part of the process, a public call was typically issued for blood donations. One of the newspapers described how "a few minutes after the media announced the bombing incident in Tel Aviv, hundreds of volunteers began swarming to the blood bank to donate."

Recovery

In a story published a week after the bombers struck Tel Aviv once again, the *Jerusalem Post* (30.3.97, p. 3) claimed that "as they gathered in cafés and restaurants Friday and yesterday, Tel Aviv residents demonstrated that it would take more than rumors of a repeat terrorist attack to keep them indoors on a sunny weekend." The reference was to the bombing at the famous Café Apropos, where three customers lost their lives. The shift manager, Erez Agmon, told the *Post* that many of the establishment's regular customers were missing, though other people came to the café especially to show support. "Some people told me that they chose our café because they figured the terrorists wouldn't hit the same place two weeks in a row," he said. The body politic enters the place of the bombing in order to reassure the recovery of normal life. This was the purpose, for example, of Prime Minister Netanyahu's ceremonial walk through Jerusalem's *mahaneh yehuda*. As described by the *Jerusalem Post* (7.8.97, p. 3), on this, his first visit to the marketplace since it was ravaged by the two suicide bombers, the prime minister's behavior was entirely ritualistic: he lit memorial candles to the 13 victims and demanded that the Palestinian leaders fight terrorism. Both these acts were a part of the ritual of recovery.

Rabin's Assassination

In describing the media coverage of Rabin's death, I will depart from the practice of the previous sections. Although the three-phase ritual of loss, regathering, and recovery also played itself out in this instance (perhaps it is inevitable under such circumstances, a kind of protective mechanism of the collective psyche), I have preferred to discuss his death as a case that lends itself particularly well to a close analysis of bodyTalk as a mediator between the physical body and the territory of the body politic. Furthermore, Rabin was perceived as a personification of the chosen body.[4]

Rabin's assassination was a formidable event in Israeli life, no less traumatic than John F. Kennedy's assassination was for Americans. One way of approaching the event from an analytical point of view would be to note that, whereas Rabin's physical body indubitably died, his social death was postponed. Rabin was kept symbolically alive in order to preserve the unity of the body politic. Amid the total confusion that struck the media at the moment of the assassination, the prime minister's body temporarily disappeared. The grand finale of the peace rally where he was struck down, with both Rabin and Shimon Peres joining in the exhilarated chanting of the "Song for Peace," projected the two leaders as part of an indivisible collective body. That is how the media wanted us to remember Rabin. In fact, Rabin's burial was never shown on television; the coverage centered instead on three places of commemoration where Rabin the symbol was kept symbolically alive: his family home; the Tel Aviv square where the murder was committed; and his grave at Mount Herzl. These three sites of public mourning drew hundred of thousands of worshippers in a threefold pilgrimage that was shown religiously on national TV.

Following the assassination, the blow to the nation and the collective body was the first to be spelled out. David Grossman, the Israeli novelist, wrote that "the epoch of the sabra has been terminated with two shots. . . . Rabin was the DNA sequence of Israel. When he died, we died too." Yossi Sarid, a member of the Knesset, wrote that now "the disease must be diagnosed and the body recovered. . . . We want to heal the body and therefore we must have a diagnosis." Yehonatan Gefen, a noted poet, novelist, and columnist, wrote in his column that "Rabin was us: one of the people, finally someone who looked like us. He was a fighter who spoke in the language of the homeland, which is the language *of the body* [italics in the original] and our mother tongue. Rabin was us, a son of a country that is the land of milk and honey and that eats its dwellers. Each one who secretly or openly spilled the poison is responsible for killing our leader." Nature, the territory, and the public mood became one in an act that was identified throughout Israel as the "earthquake." "Three shots from one of us brutally broke down this country's youth—now it seems that we have become old all at once. The country, like Rabin, did not manage to proceed naturally from youth to adulthood and old age. The brute hand that took Rabin's life threatens to take it all, take the country's life, spoil the texture of living in the State of Israel. The body of the murderer is in jail, but his evil spirit still reigns outside. We must let the nation's wounds heal." Yisrael Harel, chair of the Yesha organization, wrote that "the bullets that penetrated Rabin penetrated the body of this whole country, as well as the idea of the whole country [*shlemut ha'aretz*] and our settlement feat" (Yediot, 11.6.95, p. 15). It is noteworthy that this remark made by a devoted rightist (Yesha is an extremist group representing the settlers in the occupied territories) employs the bodyTalk rhetoric common to all the other writers quoted. That the body of the leader stands for the territory is a well-established tradition in the history of human societies. Medieval kings, for example, embodied their territories; a barren queen would be the cause of a drought that struck the country. Although modernity (not to mention postmodernism) was expected to do away with this myth, in Israel it seems to have grown stronger. The public mourning of Rabin, in the weeks following his death, was mediated mainly through the "the language of the homeland, which is the language *of the body*" (Gefen's words).

The preceding quotes were taken from the secular media largely identified with the Labor party and its leader, Rabin. As one might expect, the religious media, which are known to put a notable emphasis on bodyTalk, much more than their mainstream counterparts, also made extensive use of it in the wake of Rabin's death. Consider, for example, this comment by Yisrael Eichler in one of the right-wing organs a few years later:

Today [1997] the people understand that those who want to tear apart the Jewish people from within are more dangerous than the Arab enemy. The Arab enemy wants our land, the body, and not the mind. . . . The Arab enemy makes us more cohesive, while the inner enemy tosses its venom from the inside and dismantles the very foundations of the existence of the Jewish people. . . . Only war sacrifices (of orthodox Jews, God forbids [*rachmana litzlan*]) would re-establish the solidarity of the Israeli people, whose honor [would be] trodden on and whose blood [would be] spilt in the streets of Tel Aviv and in the suburbs, settlements [*moshavim*], and kibbutzim. The hostile media sow poison in the heart against Jews. . . . Those who follow the blood-dripping reports understand how much they [the mainstream media] take pleasure in it and make their living off others' pain. . . . Yigal Amir, a soldier, an infantry veteran [*boger Golani*], the best friend of the Shabak messenger Avishay Raviv and follower of the faith in true Zionism [*Ha'tzionut HaShlema*] opened the gash [*mursa*] two years ago, and all the puss [*mugla*] has burst out. Now it seeks a crack through which to penetrate and contaminate the whole body. . . . It seems that all the roads in the media are blocked by hearts of stone. (*Hamachane Hacharedi*, 22.11.97)

This statement, which might seem unique in its florid rhetoric to outside observers, is actually quite typical of the religious discourse. I have chosen it because it encompasses, in the scope of a paragraph, many issues that are central to the contested Israeli-Jewish collective identity. It was written amidst the debate over a Labor-initiated bill mandating the full conscription of *yeshiva-bucher* (orthodox students). Eichler's immediate target is the secular media, which he condemns for supporting the bill. The bill was promoted under the slogan "One People, One Conscription" (Am Echad, Gius Echad). The whole debate attested to the significance of military service as a defining force of Israeli collective identity. For the orthodox, it symbolized the battle between tradition and secularism, the chosen people and the people of the book vs. the chosen body and the military. The debate showed how the body, not the mind, has become the basic common denominator of Israeli contemporary identity. By enforcing mandatory conscription on orthodox Jews, the rest of the Israelis were symbolically saying, we can't have your minds, but we won't give up your bodies.

For Eichler, then, the real enemy was the "inner enemy" (the Labor party, Ehud Barak, and the non-orthodox hostile media promoting the bill). The immediate danger was the war sacrifices of orthodox Jews (following their enforced conscription). But the more general and profound danger was "the dismantling of the very foundations of the Jewish people." Both of these dangers are portrayed in bodily terms: the conscripted orthodox would become victims, whose blood would be spilt in the streets. Notice the shift to the passive voice; the active role here is altogether played by the secular society, those who "toss venom from the inside." The streets where the blood will be spilt are in the profane center of secular Israel: "Tel Aviv and the suburbs, settlements, and kibbutzim." As Eichler conceived it, in short, the danger was not only to one of the organs (the orthodox Jews) but also to the whole body—the whole people—as the "puss" released from the secular Jews sought "a crack through which to penetrate and contaminate the whole body."

Eichler's reference to Yigal Amir, the convicted assassin of Rabin, is also replete with bodyTalk, which is cunningly used to both legitimize and ex-communicate the killer. Amir, the assassin, is a chosen member of the secular body culture: a soldier, infantry veteran (*boger Golani*), and friend of a member of the security service (Shabak). But he is also a true Zionist. Eichler lists these "merits" in order of importance. Amir is first of all a soldier. This is a merit so far as the rest of society is concerned, but a stain in the eyes of the orthodox community. There is a double message here that Eichler's audience, the orthodox Jews, would have clearly understood. On the one hand, Eichler is saying, Amir is one of them (the secular Zionist society); he is their product, the product of military service and the Shabak's activities (it is well known that the Shabak field agent Avishay Raviv had inadvertently encouraged Amir's activities). On the other hand, Amir is a representative of "true Zionism." A purely cynical and well-devised argument is at work here that allows Eichler to at once ex-communicate Amir from "us," from the orthodox community, and construct him as the necessary evil (a killer) produced by "them," the secular society. Amir was merely an instrument. He was the scalpel that had opened the gash and allowed all the puss to burst out.[5]

6 WRITING THE BODY

THIS BOOK HAS raised what many readers will consider a provocative thesis in order to analyze sensitive issues. Parents' rejection of their newborns, the testing of soldiers, the commemoration and bereavement of fallen soldiers—these are all issues at the heart of contemporary Israeli society, as most Israeli politicians and journalists would put it. They are also issues that bring out, throw into relief, the underlying and ongoing Israeli attempt at defining a core collective identity. According to my argument, this core collective identity revolves around a root cultural idiom, the chosen body. I have tried to follow the footsteps and marks of the chosen body in this book, tracking it as an anthropologist-observer and as a participant Israeli mother and civilian. Overall, the book is dominated by the first, academic voice of the observer. In what follows I unleash my second, subjective voice, that of the participant.

My mother calls me up at Berkeley, my sabbatical exile, where I am writing the present book. I tell her about my book. "You mustn't publish it," she says. Silence. "It will please our enemies." When I don't respond, she says, "You think your university is going to like it?" I'm silent. "And the family in New York knows what you're writing about?" Silence. "I warn you, Jews will not tolerate hearing such bad things about Israel." I am trying to explain, validate my argu-

ment. But she doesn't contradict me. "Everything you are writing about is true," she admits, "but you don't wash your dirty linen in public."

My mother faxes me media stories published by anti-Israeli people. "There is an antisemitic group in Japan that thinks like you," she writes. "Do you want to give them a weapon? What did we want from the UN, when it said that Israel is racist, if my daughter says the same? Anyway, do you think we're the only ones that do such things? Well, carry on if that's what you want. But you should know you're doing damage to our country. It can ruin our country."

At this, I turn pale. I have palpitations; I can't stand this, I admit. I'm scared. Scared not to be loved, to be condemned, by everyone, my parents, my friends, my university, my husband, my children. I want to tell the truth, but I need my family's love and acceptance, even and especially when I disobey. I feel like a traitor. I sense that I cannot separate my skin from my country's skin. It's a nightmare. The country, ruined? I have no existence without my country. I see no possibility of separating my body from Israel. Will I be able to live if my country is ruined? I recall a talk with an Israeli artist, Meir Gal, who's been living in New York for the last eight years. "Everything that happens in Israel is glued to your skin," he told me. "Here [in the United States] there's a huge space between you and Israel. But over there, everything is happening inside your skin. There's an invasion, it's under your skin."

But I'm still at Berkeley, going down the stairs with my host and friend, Nancy Scheper-Hughes, telling her that I have to finish the book here, while I'm away and distant from Israel. I add, half-humorously: "Maybe after they read it in Israel, they will not allow me to come back?"

As I sit in Berkeley writing this book, the eyes of the naked woman with a pen in her hand draws my attention. She is the cover illustration for the book Women Writing Culture, edited by Ruth Behar and Deborah Gordon (1995). I see a woman sitting, exposing her body, and writing on it. Presenting it. But also distracted by the many eyes that stare at her.

Ruth Behar's essay in the book is written as a diary. The daughter of Jewish immigrants from Cuba, and an immigrant herself, she is bothered by her parents' eyes watching her—finally feeling like an orphan. Behar writes about the "bare-breasted woman with the eyes at her back." She tells us that "when a woman sits down to write, all eyes are on her," and that "a woman sees herself being seen" (p. 2). The words bring to mind the words Sartre used, "être vue" (being seen), in connection with antisemitism; the "Jew" is a Jew as long as he or she is seen as such. The identification "Jew" is conferred by others. To Sartre, *this was the source of Jewish vulnerability in the diaspora. Women, in Behar's reflexive account, share the same predicament.*

I realize that Behar is using the image of the naked woman as a metaphor for anthropological writing and maybe even for a rewriting of anthropology. She wants to replace the "woman exposing her breasts" (imagine a National Geographic-style picture of an African native) with the anthropologist, she wants to expose those who usually hide behind the lens, the computer, or the sterility of the academic paper. In American anthropology, there has recently been a surge of works on the body, but always on the body of others, the Other's body, the native, authentic, exotic body. This is not what I am looking for. I wonder, is American anthropology really the place to look for critical models of writing that expose my vulnerability in front of the eyes of my country? I feel and accept, however, the need to bring my (and other researchers') body "back in," back into academic research.

I think, "I am a little envious of you, Ruth Behar. When I sit here, doing my sabbatical at Berkeley, I am not bothered only by my parents' watching eyes. My country's eyes are the ones watching me. I am a woman writing on the chosen body through its various parts. As I watch Israel, I see myself being seen by my country. How will it see me? As a traitor?" I think, "My field is also my home, and so my reflection is also about home. Will I, by publishing such a book, do harm to my son who is now being conscripted? I do not want to be a symbolic orphan. And if I am declared a traitor—I would feel like an orphan. Yes, that is the word. My home will declare me an orphan—without the ties, familial and national, that my mother/parents represent even as she/they are still alive."

My argument in favor of a central cultural script that underlies Israeli society as a whole does not contradict other recent studies of Israeli multiculturalism and growing individualism. Those who live in Israel today feel the growing conflicts without the aid of academic research. Religious vs. secular, rich vs. poor, sectoral interests vs. national identity, globalization vs. nationalism—all these conflicts loom large in Israel at the turn of the twenty-first century. It is true that collectivism has lost much of its appeal today, especially in regard to the nation-building era. But Israeli society is still largely a collectivist one, since the identity of many of its Jewish citizens hinges on their identification with and commitment to the nation. Moreover, individualistic phenomena coexist with, and are even regulated by, the continuing search for a collective identity in Israel. Collective identity has remained significant as social glue even if collectivism has weakened and the distance between ethnic and religious/secular groups has grown. Even in a period regarded as post-Zionist, the culture of the chosen body still influences public opinion; witness, the continuing exclusion of people with disabilities and homosexuals, the abortion and rejection of children with impairments, and so on.

My analysis highlighted the underlying continuity of the chosen body as a central cultural script of collectivism, alongside changes that have led Israeli society in more heterogeneous directions. However, within the overall continuity of the chosen body, it also underwent changes. For example, the collectivist embodiments of the pioneer (halutz) and the soldier-sabra, which characterized the heyday of nation-building, have fallen from grace. Today, the Israeli obsession with fertility and reproduction finds new expression in prenatal genetic tests and other eugenic biotechnologies.

This study draws its examples of media events from recent years, and my interviews, with bereaved families and students, have also been carried out relatively recently. In terms of the data, I therefore believe that my analysis represents an up-to-date picture of Israeli society. While analyzing unsettling events like the helicopter crash, Rabin's assassination, terrorist bombings, and the Gulf War, whose menacing nature perhaps inevitably triggers "we" feelings, I have also discussed day-to-day activities such as the screening of newborns and soldiers. Through these events and activities, I have tried to shed light on the taken-for-granted militaristic script, its body imagery, and social repercussions. Furthermore, I have tried to ground my analysis in a detailed description of the various agents of normative control that work in the service of the chosen body.

In the "chosen body" book, I write on the subjugation of the private, personal body by the body politic, about contexts where the two become one, about doctors, teachers, but above all, about parents as the agents of that subjugation. I analyze the data and find myself sucked into their reality. While analyzing my data as an anthropologist, I take part in this arrangement as a parent whose son is being conscripted during her sabbatical, because the military changed my plans when it decided to conscript him earlier.

At Berkeley, I am having lunch with Nancy Scheper-Hughes. I say, "I need to finish writing the book here." She asks, "Why?," and I say, "Because only

*here, in the diaspora, can I really work." I begin to grasp that the diaspora is a
crucial state of mind for me, if I am to write. This is my most "sealed room." I
suddenly notice that I have intentionally invoked the image of rooms sealed
during the Gulf War, where the gas could not enter in the event of a missile
strike. This is a powerful image for me. The others are in their sealed rooms
during the Gulf War, but my sealed room is here, in Kroeber Hall.*

The practices and agents of the chosen body exist within seemingly disparate areas—for example, premarital or prenatal screening and soldiers'
screening. The rhetorical thread between birth, where a child's physical
impairment is sufficient cause for abortion, institutionalization, or seclusion within the home, and military service, where bodily criteria are crucial,
is unmistakable in the interviews conducted with parents. Although gender
proved to be a differentiating factor, both fathers and mothers emerged
from these interviews as, generally speaking, loyal agents of the chosen
body.

An unfortunate and inevitable outcome of militarization is bereavement
and commemoration. Parents' cooperation with taken-for-granted military
procedures and state ceremonies was examined as yet another facet of the
Israeli culture of the chosen body. Bereavement and commemoration are
appropriated by the Israeli collectivity and reproduced in public life so as to
sustain collective boundaries and reaffirm the sanctity of the homeland.
That is why, for example, a "traffic" accident such as the helicopter crash was
defined as a national tragedy, while the many more civilian victims of real,
daily traffic accidents usually go unnoticed. In Israel, private bodies are
enveloped by larger, collective bodies: the surroundings (the bus, the
marketplace, the center, and ultimately the land), the body social (the family, the community) and the body politic (the various ministries, the Knesset,
the prime minister, the president). The media coverage following such incidents was shown to follow a predictable ritual of loss, regathering, and
recovery, a ritual designed to reemphasize the collective boundaries in the
face of danger.

*On Wednesday or Thursday, I return to my room at Berkeley. The phone
rings, and I cheerfully answer. On the line is my friend Lin, eight years in
Pittsburgh. In a few months she plans to go to Israel for her son's conscription.
She says, "From your voice I understand that you still don't know." "What?" I*

ask. She says, "There has been a terrible accident. About 70 soldiers were killed in a helicopter crash." My cheerfulness is all gone. I have no TV. I don't know anything. I am still not into Berkeley's Israeli grapevine. I cannot sleep. I go to the university but feel out of focus. Nancy asks me about the party she's throwing for me. Who do I want her to invite? I don't care. I do not want to sound rude, so I tell her that a disaster took place in Israel. "Was it anybody you knew?" "I don't know the names," I reply. "So was it a traffic accident?" she asks. "Well, yes," I reply with hesitation. So, why is it a national event? True. It was a traffic accident. No, I don't know why it's a national event.

I receive e-mails from Israel on the extent of the disaster. "The rain blends with the tears, the sky is crying with us," a friend writes to me. She, like myself, didn't know any of the deceased in person, but it doesn't matter. I analyze the embodiment of the body politic (I live it), the embodiment of the landscape, of the media—I do a content analysis of the media and write about the body politic imprinted in the private body, and I live it. My inability to disassociate is part and parcel of this. My life blends with the country's life.

It was Nancy who invited me to come to Berkeley. I told her, there is one problem. I have to be extremely organized right now. My son is going to be in the army. I have to be here (in Israel) for that. And then Nancy drops a question on me, out of the blue: "Do you let your son be conscripted?" Nobody has ever asked me that question before.

The chosen body, as described in this book, is a masculine body; I therefore discussed the construction of Israeli manhood as bound to, and by, the bodily practices of soldiering, war, and the "fatherland." I also discussed the participation of Israeli women in this masculine ethos of the body, but with a different voice. The women who figure in the book fulfill a special role in the culture of the chosen body. It is the anthropological role of participant-observers. I identify with those women (mothers of combat soldiers, of fallen soldiers, mothers in general) because I am, like them, an agent of the normative expectations of my family and of Israeli society. I too am the object of normative control. I too did not rebel openly and courageously but only implicitly, through the body, by not doing what I was supposed to do. These small refusals included not going on annual trips or participating in physical activities. Active resistance that replaced the passive refusal came only later, even if it was expressed through a passive medium: writing the body.

Fanya—I write about bereaved mothers. About Fanya, a bereaved mother, a special one. Now, at Berkeley, I reread my fieldnotes from 25 years ago, when I first interviewed Fanya, and I recognize her subversive strategy for the first time, how she subverted the social script of the chosen body by focusing on the concrete body of her son. Now I am in Berkeley while Shay is going to be conscripted any day. I get a fax from Fanya. She was the first to contact me following Shay's birth. A few hours before the birth, I had defended the thesis in which Fanya figured as an informant. I knew I had to finish everything that had to do with the thesis before I gave birth. I didn't want to know the sex of my fetus. Knowing it was a male would have prohibited me from completing a master's thesis that dealt with the bereavement of fallen soldiers. Shay was born, and the midwife, who had been a student of mine, said happily, "Congratulations, Mother [Mazal tov, Mammale], we have a soldier." I felt heavyhearted. Two hours later, Fanya arrived, this petite bourgeoise from Poland who went so far as to climb over the hospital fence to see me. Only she understood my crying when she held Shay up and said, "Maybe by the time he's eighteen, they won't have to go to the military anymore."

I participate in Nancy's seminar. I talk about my urge to decontextualize evil. Shay is going to be conscripted. In Nancy's seminar at Berkeley, they discuss the death of the unprivileged in South America. I am talking about the death of the privileged. I think, what about the death of the Chosen? At one time, I was critical of Nancy's book. Her subject is M(other) love, the love of a Brazilian mother and how she differs from a North American mother. In fact, Nancy and many other anthropologists do not study mothers near them. They study Other mothers, m(others).

This criticism emerged again in one of the seminars. The papers on violence were also from other, distant places. In one of the meetings, Mark Sacks (not his real name) and I were invited to be guest speakers. Mark is a physical anthropologist and as a member of the Physicians' Association for Human Rights, worked with world-renowned scientists preoccupied with that issue. He opened his institute at Berkeley to me, introduced people at my request, provided access to valuable material.

After I concluded my talk about violence in Israel, Mark presented beautiful slides of physicians' activity for human rights in Nicaragua, Ruanda, Croatia, and other places. Later, taking lunch at the faculty club, over a glass of white wine and a big American "sub," Mark asks for my opinion on his pre-

sentation. "It was great," I say, "but what the students actually got was a very limited representation of human suffering; all the slides showed the beautiful American who went to the heart of darkness to fight for human rights, help the Other, the non-American, but what about human rights in the United States? Why does all the funding go for projects far from the borders of the United States?" "Thus represented," I suggested, "human suffering becomes inevitably identified with certain areas, while the United States is glorified as the guardian of human rights."

Mark was insulted. His definition of the Other is clear. Mine isn't. The Other is part of me. The American anthropologist does not share my dilemmas. The American anthropologist is committed to uncovering others' misdeeds, others' violence. For the Israeli anthropologist, the quest for collective identity begins in the Other within us.

Tami, my daughter, comes to visit me in Berkeley. During her first night, jetlagged and all, she goes over my material, an immense array of excerpts that takes up the entire floor. "Wow!" she tells me. "You've really got yourself some X-files here."

REFERENCE MATERIAL

NOTES

Complete authors' names, titles, and publication data are given in the References, pp. 157–74.

INTRODUCTION

1. For good discussions of the history of Zionism, see Avineri 1981; Laqueur 1972; and Vital 1975.

2. There are currently no reliable statistics concerning these figures in Israel. The quoted estimates are based on my interview with Prof. Mordechai Shochat, head of the Genetic Institute of the Rabin Medical Center, Tel Aviv, in May 2000.

3. Interview with Prof. Gideon Bach, head of the Human Genetics Dept., Hadassa Hospital, Jerusalem, 11.7.00

4. This observation is based also on an experiment conducted by Prof. Dan Grauer (Dept. of Zoology, Tel Aviv University), in which 12 couples who came to an adoption agency were asked to choose from 50 pictures of girls, 48 of which were of Caucasian girls from Russia. The other 2 were taken from the Nazi eugenics program (i.e., girls born to S.S. officers and Aryan-type German females). Almost all the couples chose the two German pictures (personal interview with Prof. Grauer, 11.5.00). Also of relevance is Dalit Kimor's TV documentary, "A Made-to-Order Baby," which aired on Israel's channel 1 in June 2000.

5. The theory of the chosen body emerged relatively late in the course of my research. I did not have anything of the sort in mind, for example, when I studied bereaved parents in the late 1970's. Similarly, there was no real connection between

that subject and the subject of my next major research project, the attitudes of parents toward appearance-impaired children, except for my personal interest in the margins of my culture and the behavior of people who found themselves in those margins, caught in an extreme situation. All my fieldnotes and interviews were recorded in Hebrew. All journal entries that appear in this book were translated from Hebrew, which is my native tongue.

6. The methodology of "multi-site ethnography" that I employ here is meant to juxtapose different practices and critical moments that constitute the cultural space of the Israeli life course.

7. Gluzman's (1997) analysis is based on the psychoanalytic-textual method developed by Jonathan Boyarin (1997).

8. I use the term "screening" throughout the book rather than the equivalent term "selection." Selection, to Jews, summons up the atrocities of the Nazi regime during the holocaust. However, it should be remembered that though the word screening has a neutral and scientific sound to it, the practices of screening signify something that is equivalent to selection.

9. Parallel arguments from different perspectives have been developed by several Israeli sociologists. See, for example, Zerubavel 1995; Nachman Ben-Yehuda 1995; Katriel 1986; and Katz & Gurevitch 1976.

10. I am grateful to the Stanford University Press reviewers, Prof. Ruth Linden and Prof. Ronit Lentin, who urged me to reframe the book by introducing my voice and personal biography into it.

CHAPTER 1

1. Many studies have been published about consumerism as normative control. See, for example, Ewen & Ewen 1982; B. Turner 1991a, b; Mennel 1991; Glassner 1989; Courtine 1993; and Grover 1989.

2. The sociological agenda concerning the body comprises, for example, reproduction, sexuality, and gender (Foucault 1980; Martin 1987, 1991; Laqueur 1990; Jaggar & Bordo 1989), the emotions (Lutz 1988; Rosaldo 1984; Lewis 1975), illness (Zola 1992; B. Turner 1994; Martin 1994), and postmodern consumer culture (Glassner 1988, 1989; B. Turner 1991a).

3. The typology of the physical, social, and political body hinges on a similar categorization offered by Scheper-Hughes & Lock 1987. These authors, however, use the term "individual body" to denote the phenomenological experience of the physical body. For an illustration of the anthropological study of the phenomenology of individual bodies, see Csordas 1993.

4. The "logocentric malestream rationality" is an expression taken from feminist writing. Logocentrism means that every ideology hinges on a language, a special vocabulary that generates a worldview. Male ideology has emphasized rationality as

a key concept at the expense of other aspects of reality, such as emotionality. The emotions have traditionally been perceived as something feminine. Women were presented (by men) as more emotional, and hence prone to hysteria. It is notable that the Greek word "hysteria" actually denotes the womb; here we have a linguistic linkage (womb = hysteria) that reflects the linkage between body and ideology.

5. Other feminist assertions of subversion include "deconstructing sociology" (Game 1991) and "jamming sociality" (Cixous 1986: 96).

6. In the Israeli context, "malestream ideology" focuses on militarism, nationalism, and collectivism, and is embodied in the chosen body. "Writing the body" therefore serves here as a subversive strategy for deconstructing this malestream ideology.

7. There are many books dealing with late capitalism; see, for example, Borgmann 1992; Smith 1991; and Harvey 1989.

8. An excellent study of the German body as a social mirror is Linke 1999. She focuses on the German body after Hitler, connecting militarism and masculinity in order to show how the (united) German body continued to provide a focus for Aryan aesthetics even when (and maybe because) Nazism became a taboo. In other words, the body offers a nondiscursive symbolism that provides a linkage to a past that must be silenced. On the other hand, Linke shows how the body also provides a means for change. For example, the German political Left made use of the naked body as an act of resistance to the fascism of the Third Reich, a usage made possible by the fact that nudity was already a part of leisure activities (Freikorpskulture = free body culture).

9. "Consumer culture" may seem endlessly open and full of opportunities (provided one can afford those opportunities), yet its seductions come in standard patterns. See, for example, Featherstone 1990 on the class structure of consumer culture; Gershuny & Jones 1987 on leisure pursuits; and Chaney 1993 on the "theater of shopping," tourism, and entertainment.

10. A concise list of the feminist stream includes Lesley Hazleton, *Israeli Women: The Reality Behind the Myths* (New York: Simon & Schuster, 1977); Natalie Rein, *Daughters of Rachel: Women in Israel* (Harmondsworth, Eng.: Penguin, 1979); Geraldine Stern, *Israeli Women Speak Out* (Philadelphia: Lippincott, 1979); Deborah Bernstein, *The Struggle for Equality: Urban Women Workers in Prestate Israeli Society* (New York: Praeger, 1987); Beata Lipman, *The Embattled Land: Jewish and Palestinian Women Talk About Their Lives* (London: Pandora, 1988); and Lisa Gilad, *Ginger and Salt: Yemini Jewish Women in an Israeli Town* (Boulder, Colo.: Westview, 1989).

11. See Ram 1993a for a discussion of these feminist critiques and their three typically Western thrusts: liberal, Marxist, and radical.

12. This assertion is not new. For complementary historical interpretations of this argument, see Berkowitz 1993, 1997; J. Boyarin 1997; Breitman 1988; Gilman 1991; and Mosse 1996.

13. The concept of "deep structure" is taken from Chomsky's linguistic theory. Chomsky claims that the human brain is equipped with some sort of "generative grammar" that enables it to construct sentences by connecting certain syntactic patterns. These patterns constitute the deep structure of language. The chosen body is, by way of analogy, a deep structure in the syntax of Israeli culture.

14. Ahad Ha'am (literally, "One of the People") was the pseudonym of Asher Ginsberg (1856–1927). His so-called spiritual Zionism came to be seen as an antithesis to Herzl's political Zionism (see Avineri 1981: 112–25).

15. One group that took the Zionist idea of the "new people" to its logical extreme was the Canaanite movement, whose few but ideologically minded members represented one of the more interesting (yet relatively unstudied) offshoots of Zionism (see Gonen 1975: 295–300; and Shavit 1984). In simple terms, they claimed that there had never been a Jewish nation, only a Hebrew nation. Judaism was a "malady" that spread after the destruction of the Second Commonwealth, which grew out of the misconception that a nation can be bound together by religion alone. The Canaanites believed that a Jewish nation could be brought about only by a return to nature, the land, and the Hebrew language, all symbolized in the body. The group, whose members actually called themselves Young Hebrews, got the Canaanite name from their alleged reenacting of ancient rituals. There were rumors, for example, concerning ritual dancing at night around a fire, and a naked woman who represented the goddess Ashtoret. Never more than a small minority, the Canaanites disappeared by the 1960's, but they had an important ideological impact on Israeli society.

16. The ambivalent articulation of an alternative image to that of the lamb to slaughter—namely, the image of the Jew as rebel (for instance, in identifying the Israeli national memorial day for the holocaust as also "in memory of the ghettos' uprising," and Yad Va-Shem Institute as "the memorial agency for the holocaust and heroism")—attests to the dominance of that metaphor during the first decades after the holocaust.

CHAPTER 2

1. Another American author who deals with the Jewish body is Paul Breines, whose book *Tough Jews* (1990) deals with the attempts of American Jews to distance themselves from the stereotype of the weak and effeminate Jew and find a place for themselves in the American culture of body-building.

2. It is interesting to note a phonetic linkage in Hebrew between youthfulness and chosen-ness. *Bacharut* (youth) is a modern-day Hebrew noun derived from the more traditional *bachur* (young man). Bacharut was primarily used by ideological leaders in the prestate era to denote the young generation of their movement. For example, the prestate young political cadres of one wing of the kibbutzim movement

were called Ha'Bacharut Ha'Sotzialistit (The Socialist Youth). Bacharut is probably a coined term, since to Israeli ears it sounds very much like *nivchar,* chosen. The phonetic similarity that associates youth and chosen-ness has been reproduced through a long and painful militaristic history of young soldiers who "gave their life on the altar of the state," as the common commemorative phrase goes.

3. The phenomenon of selective DI sperm banks is not unique to Israel, of course. Some American banks allow recipients to select the donor themselves, and others offer "a personal donor" program (e.g., the Xytex Corp. in Georgia, the Berkeley Sperm Bank in California). These new policies bestow DI recipients with an unprecedented ability to modify the procreative process.

4. It is important to note that the preferences found (through questionnaires) are hypothetical, because there is no choice in Israel; donor matching is performed exclusively by doctors.

5. The Zionist movement sparked a general interest in a new demographic definition of the Jewish national collectivity. That issue was probably also behind some of the work of 20th-century Jewish physical anthropologists on the characteristics of the Jewish body (see Goldberg & Abuhav 2000).

6. My research was based on observations and interviews with parents of children hospitalized in nine wards of three Israeli hospitals. The observations were conducted in the hospitals themselves, except for some 200 I made in the subjects' homes. The children observed (newborns, toddlers, and older children) suffered from internal and external deformities and various sorts of diseases. All together, I observed 100 mothers who gave birth to normal children, and the parents of 350 impaired (appearance or otherwise) newborns. The 350 newborns represented all the impaired children born in the three hospitals during my period of observation; 250 of them suffered from impaired appearance (e.g., spina bifida, cleft lip, bone malformations), and 100 from internal abnormalities (e.g., heart or kidney diseases). I also collected data on the parents of 50 premature babies, all the premature births in one of the hospitals. For a detailed description of this project, see Weiss 1994.

7. Personal communication with the administrators of medical centers for people with impairments in Oakland, Calif., and New York.

8. Instances of the rejection, abandonment, killing, and abuse of such children have been widely reported since the times of the ancient Greeks (for historical accounts, see Pollock 1983; Hunt 1972; and de Mause 1976). In Europe, the handicapped child was often considered to be a "changeling," a creature put in the cradle by fairies who stole the mother's real child. Irish changelings were often "helped" to return to the spirit world whence they came, in some cases by burning them in the family hearth (Scheper-Hughes 1991: 376). In almost all cases of handicap, mental retardation, and impaired appearance in infants, then, there is a powerful stigmatization at play. Examples abound not only in Western history but also cross-culturally. The African Nuer studied by Evans-Pritchard (1956) referred to the physically

deformed infant as a "crocodile" child, and presumably submerged it in water. A similar attitude toward "witch children" (characterized by such physical abnormalities as breech presentation, congenital deformity, and facial or dental malformation) was found until very recently among the West African Bariba (Scheper-Hughes 1991: 376). Physical abnormalities are also among the "signs" that tell Brazilian Shantytown mothers their child is "taboo," which leads to his or her mortal neglect (see ibid. for a fuller cross-cultural list of such practices).

9. Stigmatization was also shown in studies of handicapped children (Goodman et al. 1963), infants with facial deformities (Salyer et al. 1985), and amputee children (Centers & Centers 1963).

10. For an analysis of the personal, as well as territorial, stigmatization of appearance-impaired children in the home, see Weiss 1994.

11. Badinter (1995: 67) argues that in most human societies, masculine identity has to be acquired through three basic elements. The army has always been a masculine identity-processor, and all the following general elements are well present in Israeli military service (for similar analyses, see Sion 1997; and Lieblich & Perlow 1988). Badinter's first element is a critical threshold that must be crossed. At some point during his adolescence, the child must abandon his former identity in order to become a man. This involves an educational process, since becoming a man requires a voluntary effort. In our case, military conscription denotes that threshold. It marks the youth's effort to become a man. Officers commonly tell their combat soldiers during infantry basic training that "first, we must turn you from citizens into soldiers; then, we must make warriors out of you." The second element consists of tests. Manhood is won after a battle, against another man, or against oneself, and this battle often includes physical and mental pain. Military trainees are subjected to stringent physical and psychological screening. This happens in an especially intensive way during the trainees' transition period, which is basic training. The third element is the father's role. It is he who is traditionally responsible for making a man of his son. In the army, the commander-soldier dyad represents the father-son connection. During basic training, new trainees are called "wet and fresh bodies" (similar to newborns; see Sion 1997). The trainees are infantilized, become young children again as they find themselves in a world that is totally dominated and controlled by their officers in charge. Like babies, most of the trainees' pleasures are oral, such as food or cigarettes, and these are controlled by parents. Adult issues such as sex are excluded from the infantile world of the trainee.

12. Bnei Akiva (literally, "Sons of [Rabi] Akiva") is the name of a right-wing, Zionist-religious youth movement.

13. Gush Emunim (literally, "The Block of the Faithful") is a right-wing, national-religious social movement active in settlements.

14. Aggressive children were thus found to draw figures with protruding teeth, long arms, and big hands (Wysocki & Wysocki 1977); obese women drew a larger fig-

ure (Kotkov & Goodman 1953); hospitalized children drew, inside their figure drawing, the sick organ (Tait & Ascher 1955); amputated children omitted parts of the legs or distorted them (Silverstein & Robinson 1956); and children with cancer used only a small portion of the sheet, drew with many erasures, sharp lines, and no use of colors, and their figures presented internal gaps and various organs protruding out of the body (Spinetta et al 1981).

15. Jewish laws of conduct (the halacha) generally prohibit autopsies, which are regarded as compromising the "honor" (integrity) of the dead. Representatives of the Holy Society and the Army Rabbinate are present daily in the institute to ensure that it maintains the Jewish law. Autopsies are never automatic; they require permission and negotiations. Another conflict between Jewish religion and the institute has to do with the identification of the dead person. According to the halacha, visual identification of a corpse by a relative is sufficient; but in the eyes of the institute's staff, such identification is not scientifically valid.

16. The policing of "healthy" sexuality is of course not limited to forensic physicians, but is seen as a function of the medical profession as a whole. Consider, for example, the medical establishment's initial reaction to contraception. According to Mosse (1996: 33), the British medical society's journal, Lancet, opposed contraception when its use was debated in 1870 because "a nation fruitful in healthy organisms must, in the struggle for existence, displace and swallow up a nation that is abandoned to conjugal onanism."

17. Circumcision is the most "popular" religious practice observed in Israel (by about 98% of the Jewish population, including even people who consider themselves anti-religious). The tension between the orthodox and the secular hence disappears in this practice, which is nevertheless one of the most intrusive, violent, and irreversible. (I am grateful to Tamir Sorek for this observation.) In my terms, this practice exemplifies the imprinting of the collective body onto the personal body. Furthermore, new immigrants from countries where circumcision is not practiced (e.g., the former USSR) are circumcised in Israel with the encouragement of the state.

18. In fact, soldiers' bodies are hardly ever used for organ harvesting. Not only in this, but in other ways, Israel is unique in the trafficking of organs. It buys the largest number of organs, proportionally, in the global market, but the domestic rate of organ harvesting is very low, and no organs are ever sold to the global market. For Scheper-Hughes 2000, the religious-ethnocentric position expressed by Israel in this regard is yet another indication of the existence of the culture of the chosen body.

19. The goal of the genetic test was to confirm beyond doubt whether the graves contained the remains of Yemenite children who died from medical problems (as the establishment said) or not.

20. The Yemenite children affair illustrated the dark side of collectivism.

21. Some 57,000 Ethiopian Jews currently reside in Israel. Many of them migrat-

ed from Sudan and Ethiopia in two dramatic airlifts, the first of which was called "Operation Moses." The Ethiopians in Israel are stigmatized because of their black skin, different culture and language, "uncertified" religious status (in the eyes of the orthodox establishment), and poor living conditions. See Kaplan 1992; Salamon 1995; and Wagaw 1993.

CHAPTER 3

1. The tragic crash also raised a limited discussion on the need to pull out of south Lebanon. But this matter, however relevant, was pushed aside by the overall "moment of solidarity." It was given more publicity by foreign media covering the Israeli scene than by the local media.

2. See, for example, Goody 1962; Danforth 1982; L. Taylor 1989; Aries 1974, 1981; and Palgi & Abramovitz 1984.

3. Sociologists, historians, and anthropologists have demonstrated the existence of this "national hero system" in various commemorative settings, such as military cemeteries (Bloch 1971; Mosse 1979), war memorials (Griswold 1986; McIntyre 1990), and mourning rites (Schwartz 1982; Robin et al. 1991).

4. The areas of study of this new social inquiry include the use of the fallen as a national symbol in Hebrew war literature (Miron 1992), in political propaganda (Gertz 1984), in the design of war memorials (Levinger 1993; Azaryahu 1992), in the life of war widows (Shamgar-Handelman 1986), and in the psychological treatment of the bereaved (Palgi & Durban 1995; Witztum & Malkinson 1993).

5. All three days have been the subject of sociological analysis, a fact that attests to their importance in Israeli public life. On their history and meaning, see Young 1990 (Holocaust Day); Handelman & Katz 1990 (Memorial Day); and Don-Yehiya 1988 (Independence Day). Bet-El & Ben Amos 1993 treat the Holocaust and Memorial days together.

6. Although ultra-orthodox schools are partially subsidized by the state and so formally pay lip service to the MDF, they apparently do not perform the Ministry of Education's prescribed ceremony. A few years ago, there was a public uproar when telecasts showed ultra-religious pupils failing to stand still for a moment of silence at the sound of the MDF siren. This rekindled the traditional debate on why the state should subsidize or even recognize the ultra-religious if they are exempt from military service.

7. Iyar is the seventh Jewish month, which usually falls around April–May. The fifth of Iyar (Hey Be'Iyar) is the Jewish date of the declaration of independence, immediately followed by the War of Independence, 1948.

8. The translation is from a bilingual brochure, *Ceremony for Yom Hazikaron—Memorial Day for Israel's Fallen*, an anthology of selected readings compiled and distributed by the Ministry of Foreign Affairs through its embassies and consulates. The

names of the translators are not given in the brochure, from which the following texts are also quoted.

9. *The Bath Queen* (1970) was Levin's third political satire (after *You, Me and the Next War* and *Ketchup,* written in 1968–69). The play criticizes the national cult of the fallen, which emanated, according to Levin, from the 1967 war.

10. The Keren Kayemet (the Jewish National Fund) was established in 1901 to purchase land in Palestine (and Syria, at that time). The now famous "blue box" was placed in many Jewish community centers around the world, as well as in Jewish schools in Palestine, to encourage donations to the fund (Gvati 1985: 39).

11. As the name of the H&N exhibit implies, it was not merely about bereavement but also about photography, or more precisely about the symmetrical ideologies of hegemonic representation/reproduction embodied in those two practices. Ariela Azoulay, the exhibit's curator, wrote in the catalog that "H&N are two photographers who have constructed a confession cell for themselves so that they can mourn the death of photography." The confession cell was constructed in the exhibit as an analogy to the photographic cell—the camera obscura—and the act of photography was compared with the act of mourning through their symmetrical commemorative power.

12. The confession cell could also be read as an allusion to the process of subjugation described by Michel Foucault (1980). He cites as an example the way in which the Church turns a believer into a "confessional animal," a *homo docilis* who regards confession as an inner demand, and not the effect of a power that constrains him. Such a successful internalization results in a genuine need of the subject to fulfill the "requests" of the collective, which is, according to Foucault, normative control in its most efficient (modern) form.

13. In February 1992, H&N were involved in a group exhibition called "Olive Green," which was shown at (the same) Bugrashov Gallery in Tel Aviv and curated by (the same) Ariela Azoulay. In Israel, olive green is the color of the army uniform (equivalent to "olive drab" in English). In her catalog, the curator, who initiated the exhibit by suggesting its organizing theme, analyzes H&N's works in a manner that suits their 1993 exhibit just as well: "The works of Erez Harodi and Nir Nader [present] the transition from the political object to the gaze that becomes politicized. . . . Nader and Harodi bury the anonymous soldier and with him the anonymous viewer, wrapped in olive-green shrouds. Everyone looks at [at the viewer], manipulates him . . . and he who was once an anonymous viewer in all his glory, the ideal viewer, is subjected to the gaze of another, the 'gaze' of space, a gaze produced by the very organization of the space that is gloriously built, and contains many sections, guardhouses, pits, and isolation cells. School and army, family and state, private and public spaces, all are made of cells upon cells, niches upon niches, into which it is necessary to pack people, to call the body to order, to train them and demand the fitting of their body to the design of the cell, the design that is given in advance and is iden-

tical for all. A design that is meant to emplace the identical, to spew out the different who refuses to erase his differences, to straighten the corners and become identical, and in brief to engulf all. . . . From the position of an artist who stands guard by means of his work, Nader and Harodi try to jumble the power relations . . . such as soldier vs. an order, citizen vs. an order, soldier obeying, soldier refusing to obey" (pp. 33–35).

14. Israel's approach to the disabled should not be compared to that of other countries only in legal terms. For example, there is little of the public discourse on disability rights that has come to consume American society. On the U.S. discourse, see Gelya Frank 2000.

CHAPTER 4

1. Max Nordau, "Zionist Writings," vol. A (1936), tr. Y. Yevin and H. Goldberg, in the Zionist Library, Jerusalem. There is a long antisemitic tradition identifying Jews with femininity. As Gluzman 1995 notes, Jung wrote that "Jews and women are alike: since they are both physically weak, they must aim at the breaches in the armor of their rivals"; Freud similarly described the Jewish mentality in discussing circumcision; and Charcot, the Parisian physician who was an authority on hysteria (and who oversaw Nordau's medical training), thought that Jews had a special tendency to hysteria. On the connection between the woman and the Jew in Nordau's thought, see Gilman 1993.

2. The woman poet Bat-Sheva Altshuler, who participated in the early fighting as a combatant, focuses on the dead body in her poems.

3. Another contemporary gendered narrative that relates to sacrificial bodies was the mothers' protest movements at the sacrifice of their sons in Lebanon, beginning with Mothers Against Silence after the 1982 invasion. This and later groups (like the Four Mothers organization) appear at first blush as feminine loci of resistance against the standardized script of soldiering and militarization. However, a closer look reveals that they actually accepted the hegemony of the chosen body. Based on interviews with some of the activists in the Four Mothers group, I argue that they in fact cooperated with the military and the political establishment. The concern for the children's safety reflected a motherly discourse of tribal protection, rather than some universal morality or pacifism. The "mother of a paratrooper," in a famous public letter to her son's commander, called on him to protect her son even at the cost of injuring Palestinian children (Atzmon 1997: 121). These grass-roots protest movements were not aimed at the state's conscription policy, the legitimacy of the IDF, or the practices of soldiering. They were pointed specifically at the military presence in Lebanon and the risk to their children's life. All of them consistently argued in favor of active military service, and the mothers were proud to describe the ordeals their sons had to go through to get into the elite units in which they were

now serving. In 1993, the Labor government ordered a unilateral withdrawal of the IDF from Lebanon. The position of the protesters therefore became part of the establishment, basically because it never really threatened the ideology of the chosen body. Only one women's protest group, New Profile, has defied the premises of the collectivity, but its size and influence are negligible.

4. Gibush is a concentrated training during which youngsters who wish to join volunteer units are tested for their suitability. Thus, "passing the gibush" indicates acceptance into the unit of choice.

5. Atrophine was supplied to the Israeli public in the Gulf War as an antidote to nerve gas.

6. For analyses of international attitudes toward the Gulf War, see Kellner 1992; MacArthur 1992; Mowlana et al. 1992; and P. Taylor 1992.

7. One important example is Chomsky's article "What War?" Answering his own question, Chomsky (1992: 51), wrote: "As I understand the concept 'War,' it involves two sides in combat, say, shooting at each other. This did not happen in the Gulf."

8. I did interview six doctors, including four men. All six discussed the war effort in more purely professional terms than the nurses, as something to be faced like any other calamity. This attitude did not change before, during, or after the war, and was the stance of the female doctors as well as the males. All of them told me that they (and their families) were already used to emergency shifts and terrible work hours at the expense of family life. In three cases, the respondent's spouse was also a doctor.

9. For example, in one of the rare public documentaries on the Gulf War, "Front Line at Home" (a video distributed by the Electronic Communication Division of the World Zionist Organization in 1994), there is no mention of the nurses.

CHAPTER 5

1. For empirical studies, see Dobkin 1992; Paletz & Schmid 1992; and Wittebols 1992. Working under the same thesis of "domination through fear," or the conscription of the media for national security goals, Gabriel Weimann (1990) found a similar case in Israel, where he sees the media functions as status-conferral and agenda-setting.

2. All of the material on the reportage is drawn from my collection of obituaries, newspaper articles, and TV transcripts (published in Weiss 1997). My study of the Israeli press over a 20-year period reveals a significant use of the body as a rhetorical device for references to the collectivity. This rhetoric was once particularly characteristic of the right wing; recently it has become a trend that crosses political camps. Its use is augmented in times of national trouble and threat, such as suicide bombings, Intifada, and wars. It is interesting to note that the rhetoric of the U.S. media following the September 11 tragedy was quite different. As it happened, I

was in Manhattan during that time of national catastrophe, and in none of the accounts that I read or heard was there any rhetorical reference to the "body of the nation," despite a considerable increase in the nationalist-religious discourse. This observation strengthens my argument that the chosen body and its rhetorical devices are a cultural script unique to Israeli society.

3. This autopsy did not go against the general policy of the institute not to touch fallen soldiers' bodies. As I mentioned earlier, the institute does not harvest skin or use the corpses for medical practicing. But autopsies are conducted when required by law or the military. In this case, it was considered necessary to understand what had caused the accident, and to test the hypothesis that the pilot might have had a stroke before the crash.

4. An indication of the importance of Rabin's body is found in the interviews I conducted with the staff of the Institute of Forensic Medicine. One of the assistant pathologists told me that it was an autopsy he would never forget, and not just because it was the Israeli premier. The assistant noted that what impressed him were the heart and the lungs. "A man who smoked so much—His organs were that healthy. A man who was 73 and had the whole world on his shoulders, and yet was so healthy." The assistant added that, contrary to standard procedure, where the internal organs are taken out of the body and spread on the table, the autopsy was conducted without removing any internal organ. This extraordinary fact can be explained by the desire to keep the chosen body complete and intact. On leaving the autopsy room, Professor Hiss (who was in charge of the operation) described to me how he "expected the sun not to be shining outside." He was astonished to see that the sun was shining. Here we see the anticipation that nature would stop its course and take part in the national mourning for the chosen body.

5. On February 21, 2002, the High Court of Justice decided to prolong the exemption of the orthodox from military service, thus maintaining the status quo.

REFERENCES

Abeliovich D., A. Quint, N. Weinberg, G. Verchezon, I. Lerer, J. Eckstein, and E. Rubinstein. 1996. "Cystic Fibrosis Heterozygote Screening in the Orthodox Community of Ashkenazi Jews: The Dor Yesharim Approach and Heterozygote Frequency," *European Journal of Human Genetics*, 4.

Abdo, Nahla, and Nira Yuval-Davis. 1995. "Palestine, Israel and the Zionist Settler Project," in *Unsettling Settler Societies: Articulations of Gender, Race, Ethnicity and Class*, ed. D. Stasiulis and N. Yuval-Davis, pp. 291–322. London: Sage.

Alexander, David. 1985. *The Jester and the King: Political Satire in Israel.* Tel Aviv: Sifriyat Poalim (in Hebrew).

Almog, Oz. 1994. "The Sabra—A Sociological Profile." Ph.D. diss., Dept. of Sociology and Anthropology, Haifa University (in Hebrew).

———. 1997. *The Sabra— A Profile.* Tel Aviv: Am Oved (in Hebrew).

Amir, D., and O. Binyamini. 1992a. "Abortion Approval as Ritual of Symbolic Control," in *The Criminalization of Women's Body*, ed. C. Feinman, pp. 1–25. Binghampton, N.Y.: Haworth Press.

———. 1992b. "The Abortion Committees: Educating and Controlling Women," *Journal of Women and Criminal Justice*, 3: 5–2

Amir, D., and D. Navon. 1989. "The Politics of Abortion in Israel." Unpublished paper in the Sapir Center, University of Tel Aviv (in Hebrew).

Amnesty International. 1989. *Guatemala: Human Rights Violations.* London: AMR.

———. 1990. *Guatemala: Extra-Judicial Executions.* London: AMR.

Aries, P. 1974. *Western Attitudes Toward Death.* Baltimore, Md.: Johns Hopkins University Press.

————. 1981. *The Hour of Our Death*. New York: Vintage Books.

Arkin, W., and L. R. Dubrofski. 1978. "Military Socialization and Masculinity," *Journal of Social Issues*, 34, 1: 151–68.

Aronoff, Myron. 1989. *Israeli Visions and Divisions*. New Brunswick, N.J.: Transaction.

————. 1993. "The Origins of Israeli Political Culture," in *Israeli Democracy Under Stress*, ed. E. Sprinzak and L. Diamond, pp. 47–63. Boulder, Colo.: Lynne Rienner.

Asad, T. 1997. "On Torture, or Cruel, Inhuman and Degrading Treatment," in *Human Rights, Culture and Context: Anthropological Perspectives*, ed. R. Wilson, pp. 111–34. London: Pluto Press.

Atzmon, Yael. 1997. "War, Mothers, and a Girl with Braids," *Israel Social Science Review*, 12, 1: 109–28.

Avineri, Shlomo. 1981. *The Making of Modern Zionism: The Intellectual Origins of the Jewish State*. New York: Basic Books.

Azaryahu, Maoz. 1992. "War Memorial and the Commemoration of the Israeli War of Independence, 1948–1956," *Studies in Zionism*, 13, 1: 57–77.

Badinter, Elizabeth. 1981. *The Myth of Motherhood*. Souvenir Press: London.

————. 1995. *On Masculine Identity*. New York: Columbia University Press.

Bar Yosef, R., and D. Padan-Eisenstark. 1975. *Men and Women in War: Changes in the Role System Under Stress*. Jerusalem: Labor and Welfare Research Institute, Hebrew University (in Hebrew).

Baudrillard, Jean. 1991. *La Guerre du Golf n'a pas eu lieu*. Paris: Galilee.

Bauman, Zygmunt. 1992. *Mortality and Immortality and Other Life Strategies*. Cambridge, Eng.: Polity Press.

Becker, Ernst. 1971. "The Relativity of Hero-Systems," in Becker, *The Birth and Death of Meaning*, chap. 10. New York: Free Press.

Bellah, R., R. Madsen, W. Sullivan, A. Swidler, and S. Tipton. 1985. *Habits of the Heart: Individualism and Commitment in American Life*. New York: Harper & Row.

Ben-Ari, Eyal. 1989. "Masks and Soldiering: The Israeli Army and the Palestinian Uprising," *Cultural Anthropology*, 44: 372–89.

————. n.d. *Mastering Soldiers: Conflict, Emotions and the Enemy in an Israeli Military Unit*. Forthcoming.

Ben-David, A., and Y. Lavee. 1992. "Families in the Sealed Room: Interaction Patterns of Israeli Families During SCUD Missile Attacks, *Family Process*, 31: 35–44.

Ben-David, Joseph. 1972. "Professionals and Unions in Israel," in *Medical Men and Their Work*, ed. E. Freidson and J. Lorber. NewYork: Aldine Alberton.

Ben-Eliezer, Uri. 1988. "Militarism, Status and Politics," Ph.D. diss., Tel Aviv University (in Hebrew).

————. 1995. *The Emergence of Israeli Militarism, 1936–1956*. Tel Aviv: Zmora Bitan (in Hebrew).

Ben-Ezer, Ehud. 1967. "The Image of 'the Jew' in the Literature of the 'Native-Born,'" *Ot*, pp. 107–17 (in Hebrew).

Ben-Yehuda, Nachman. 1995. *The Masada Myth: Collective Memory and Mythmaking in Israel*. Madison: University of Wisconsin Press.

Ben-Yehuda, Netiva. 1981. *1948—Between Calendars*. Jerusalem: Keter (in Hebrew).

Berger, Arthur. 1996. *Manufacturing Desire: Media, Popular Culture, and Everyday Life*. New Brunswick, N.J.: Transaction.

Berkowitz, Michael. 1993. *Zionist Culture and West European Jews Before the First World War*. Cambridge: Cambridge University Press.

———. 1997. *Western Jewry and the Zionist Project, 1914–1933*. Cambridge: Cambridge University Press.

Berthelot, J. M. 1986. "Sociological Discourse and the Body," *Theory, Culture and Society*, 3: 155–64.

Best, J. 1989. *Images of Issues: Typifying Contemporary Social Problems*. New York: Aldine de Gruyter.

Bet-El, I., and A. Ben-Amos. 1993. "Rituals of Democracy: Ceremonies of Commemoration in Israeli Schools," *Neue Sammlung*, pp. 52–58.

Beuf, A. H. 1990. *Beauty Is the Beast: Appearance-Impaired Children in America*. Philadelphia: University of Pennsylvania Press.

Biale, David. 1986. *Power and Powerlessness in Jewish History*. New York: Shocken Books.

———. 1992a. *Eros and the Jews*. New York: Basic Books.

———. 1992b. "Zionism as an Erotic Revolution," in *Eros and the Jews*, pp. 176–204.

Bilu, Yoram, and Eyal Ben-Ari. 1992. "The Making of Modern Saints: Manufactured Charisma and the Abu-Hatseiras of Israel," *American Ethnologist*, 19, 4: 672–88.

Birenbaum-Carmeli, Daphna, and Yoram S. Carmeli. 2000. "Physiognomy, Familism and Consumerism: Preferences Among Jewish Israeli Recipients of Donor Insemination." Unpublished manuscript.

Bloch, M. 1971. *Placing the Dead*. London: Seminar Press.

Bloom, A. R. 1982. "Israel: The Longest War," in *Female Soldiers*, ed. N. Goldman, pp. 137–65. Westport, Conn.: Greenwood Press.

Borg, A. 1991. *War Memorials: From Antiquity to the Present*. London: Leo Cooper.

Borgmann, A. 1992. *Crossing the Postmodern Divide*. Chicago: University of Chicago Press.

Boswell, J. 1988. *The Kindness of Strangers: The Abandonment of Children in Western Europe from Late Antiquity to the Renaissance*. New York: Pantheon.

Bourdieu, P. 1977. *Outline of a Theory of Practice*. Cambridge: Cambridge University Press.

Boyarin, Daniel. 1993. *Carnal Israel: Reading Sex in Talmudic Culture*. Berkeley: University of California Press.

————. 1997. "The Colonialist Carnival: Zionism, Gender, Mimesis," *Teoria Ubi-koret*, 11: 123–44.

Boyarin, Jonathan. 1997. *Palestine and Jewish History: Criticisms at the Borders of Ethnography.* Minneapolis: University of Minnesota Press.

Braidoti, Rosi. 1994. *Nomadic Subjects: Embodiment and Sexual Difference in Contemporary Feminist Theory.* New York: Columbia University Press.

Breitman, Barbara. 1988. "Lifting Up the Shadow of Anti-Semitism: Jewish Masculinity in a New Light," in *A Mensch Among Men: Explorations in Jewish Masculinity,* ed. H. Brod, pp. 101-17. Freedom, Calif.: Crossing Press.

Buber, Martin. 1961. *Ways in Utopia.* Tel Aviv: Am Oved (in Hebrew).

Butler, Judith. 1990. *Gender Trouble: Feminism and the Subversion of Identity.* New York: Routledge.

Campbell, D. 1993. "Women in Combat: The World War II Experience in the US, GB, Germany and the Soviet Union," *Journal of Military History,* 57: 301–23.

Canetti, Elias. 1973. *Crowds and Power.* Harmondsworth, Eng.: Penguin.

Carmeli, Yoram S., Daphna Birenbaum-Carmeli, Yigal Madjar, and Ruth Weissenberg. 2000. "Hegemony and Homogeneity: Donor Choices of Israeli Recipients of Donor Insemination." Unpublished manuscript, in Department of Sociology and Anthropology, Haifa University.

Centers, L., and R. Centers. 1963. "Peer Group Attitudes Towards the Amputee Child," *Journal of Social Psychology,* 61: 127–32.

Chaney, David. 1993. *Fictions of Collective Life.* London: Routledge.

Chomsky, Noam. 1989. *Necessary Illusions: Thought Control in Democratic Societies.* Boston: South End Press.

————. 1992. "What War?," in *Triumph of the Image: The Media's War in the Persian Gulf—A Global Perspective,* ed. H. Mowlana et al., pp. 51-63. Boulder, Colo.: Westview.

Cohen, Irun R. 1992. "The Cognitive Principle Challenges Clonal Screening," *Immunology Today,* 13, 11: 441–50.

Cohen, S. 1997. "Towards a New Portrait of the (New) Israeli Soldier," *Israeli Affairs,* 3, 3-4: 77–117.

Connell, R. W. 1987. *Gender and Power.* Stanford, Calif.: Stanford University Press.

————. 1995. *Masculinities.* London: Polity Press.

Cooke, M. 1993. "Wo-Man: Retelling the War Myth, in *Gendering War Talk,* ed. M. Cooke and A. Woollacott, pp. 177–205. Princeton, N.J.: Princeton University Press.

Courtine, J.-J. 1993. "Les Stakhanovistes du narcissisme," in *Le Gouvernement du corps* (special issue of *Communications,* 56), pp. 225–35. Paris: Seuil.

Csordas, T. 1993. "Somatic Modes of Attention," *Cultural Anthropology,* 8, 2: 135–56.

Danet, Brenda, Yosefa Loshitzky, and Haya Bechar-Israeli. 1993. "Masking the Mask:

An Israeli Response to the Threat of Chemical Warfare," *Visual Anthropology*, 6: 229–70.

Danforth, L. 1982. *The Death Ritual of Rural Greece*. Princeton, N.J.: Princeton University Press.

Dobkin, Bethami. 1992. *Tales of Terror: Television News and the Construction of the Terrorist Threat*. New York: Praeger.

Don-Yehia, E. 1988. "Festival and Political Culture: Independence Day Celebrations," *Jerusalem Quarterly*, 45: 61–84.

Doner, Batya. 1991. *Real Time: The Gulf War—Graphic Texts*. Exhibition catalog, Tel Aviv Museum of Art, no. 4/91.

Douglas, Mary. 1966. *Purity and Danger*. Harmondsworth, Eng.: Penguin Books.

———. 1973. *Natural Symbols*. New York: Pantheon.

Douglas, M., and M. Calvez. 1990. "The Self as Risk Taker: A Cultural Theory of Contagion in Relation to AIDS," *Sociological Review*, 38, 3: 445–563.

Eilam, Yigal. 1973. *The Hebrew Regiments in the First World War*. Tel Aviv: Ma'arachot (in Hebrew).

Eilberg Schwartz, Howard, ed. 1992. *People of the Body: Jews and Judaism from an Embodied Perspective*. SUNY series *The Body in Culture, History, and Religion*. Albany: State University of New York Press.

Eisenstadt, S. N. 1985. *The Transformations of Israeli Society: An Essay in Interpretation*. London: Weidenfeld & Nicholson.

Ehrenreich, B., and D. English. 1976. *Complaints and Disorders: The Sexual Politics of Sickness*. London: Writers and Readers Publishing Cooperative.

Elboyn-Dror, R. 1996. "The Youth Culture of the Early Immigration Waves," *Alpayim*, 12: 104–36 (in Hebrew).

Elon, Amos. 1971. *The Israelis: Founders and Sons*. New York: Holt, Rinehart & Winston.

Enloe, C. 1988. *Does Khaki Become You? The Militarization of Women's Life*. London: Pandora Press.

Evans-Pritchard, E. E. 1956. *Nuer Religion*. Oxford: Clarendon Press.

Even-Zohar, Itamar. 1981. "The Emergence of a Native-Hebrew Culture in Palestine: 1882–1948," *Studies in Zionism*, 4: 167–84.

Ewen, S., and E. Ewen. 1982. *Channels of Desire*. New York: McGraw-Hill.

Falk, Raphael. 1998. "Zionism and the Biology of the Jews," *Science in Context*, 11, 3–4: 587–607.

Featherstone, Mike. 1990. *Consumer Culture and Postmodernism*. London: Sage.

Featherstone, Mike, Mike Hepworth, and Brian Turner, eds. 1995. *The Body: Social Process and Cultural Theory*. London: Sage.

Fischer, Shlomo. 1998. "Jewish Salvational Visions, Utopias, and Attitudes Toward the Halacha," *International Journal of Comparative Sociology*, 29, 1-2: 62–75.

Florian, V., and S. Katz. 1983. "The Impact of Cultural, Ethnic and National Variables on Attitudes Towards the Disabled in Israel," *International Journal of Intercultural Relations*, 7: 167–79.

Florian, V., and E. Shurka. 1981. "Jewish and Arab Parents' Coping Patterns with Their Disabled Child in Israel," *International Journal of Rehabilitation Research*, 4, 2: 201–4.

Foucault, Michel. 1971. *Madness and Civilization: A History of Insanity in the Age of Reason.* London: Tavistock.

———. 1973. *The Birth of the Clinic: An Archeology of Medical Perception.* New York: Vintage Books.

———. 1977. *Discipline and Punish: The Birth of the Prison.* New York: Pantheon.

———. 1980. *Power/Knowledge.* Brighton, Eng.: Harvester Press.

Frank, A. 1990. "Bringing Bodies Back In: A Decade Review," *Theory, Culture and Society*, 7: 131–62.

Frank, G. 2000. *Venus on Wheels.* Berkeley: University of California Press.

Freedman, Dan, and Jacqueline Rhoads, eds. 1987. *Nurses in Vietnam: The Forgotten Veterans.* Austin: Texas Monthly Press.

Friedmann, Georges. 1967. *The End of the Jewish People?* Garden City, N.Y.: Doubleday.

Fuchs, R. 1984. *Abandoned Children: Foundlings and Child Welfare in Nineteenth-Century France.* Albany: State University of New York Press.

———. 1992. *Poor and Pregnant in Paris: Strategies for Survival in the Nineteenth Century.* New Brunswick, N.J.: Rutgers University Press.

Gal, Reuven. 1986. *A Portrait of the Israeli Soldier.* New York: Greenwood Press.

Gamarnikow, E. 1991. "Nurse or Woman: Gender and Professionalism in Reformed Nursing, 1860–1923," in *Anthropology and Nursing*, ed. P. Holden and J. Littlewood, pp. 110–29. London: Routledge.

Game, Anne. 1991. *Undoing the Social: Towards a Deconstructive Sociology.* New York: Open University Press.

Game, Anne, and R. Pringle. 1983. *Gender at Work.* Sidney: Allen & Unwin.

Gellner, Ernest. 1985. *Relativism and the Social Sciences.* Cambridge: Cambridge University Press.

Gershuny, J., and S. Jones. 1987. "The Changing Work/Leisure Balance in Britain, 1861–84," in *Sport, Leisure and Social Relations*, ed. J. Horne et al. London: Routledge.

Gertz, Nurit. 1984. "Social Myths in Literary and Political Texts," *Poetics Today*, 7, 4: 621–39.

Gerzon, Mark. 1982. *A Choice of Heroes: The Changing Face of American Manhood.* Boston: Houghton Mifflin.

Giddens, Anthony. 1991. *Modernity and Self-identity: Self and Society in the Late Modern Age.* Cambridge: Polity Press.

Gilman, Sander. 1991. *The Jew's Body.* London: Routledge.

Gilmore, David. 1990. *Manhood in the Making: Cultural Concepts of Masculinity*. New Haven, Conn.: Yale University Press.

Glassner, Barney. 1988. *Bodies: Why We Look the Way We Do and How We Feel About It*. New York: Putnam.

———. 1989. "Fitness and the Postmodern Self," *Journal of Health and Social Behavior*, 30: 180–91.

Gluzman, Michael. 1995. "Body as Text, Masculinity as Language: On Body Language in the Book of Internal Grammar," *Teoria UBikoret* (in Hebrew).

———. 1997. "The Yearning for Heterosexuality: Zionism and Sexuality in Alteneuland," *Teoria UBikoret*, 11: 145–60 (in Hebrew).

Goldberg, Harvey, and Orit Abuhav. 2000. "Race Places: Jewish Identities in Anthropological Perspectives. Aspects of Physical Anthropology in the History of Israeli Anthropology." Paper presented at the annual conference of the American Anthropological Association, San Francisco, Calif.

Goldman, N. L., ed. 1982. *Female Soldiers—Combatants or Noncombatants?* Westport, Conn.: Greenwood Press.

Goldstein, S., ed. 1980. *Law and Equality in Education*. Jerusalem: Van Leer Foundation.

Gonen, Jay Y. 1975. *A Psycho-History of Zionism*. New York: New American Library.

Goode, E., and N. Ben-Yehuda. 1994. *Moral Panics: The Social Construction of Deviance*. Oxford: Blackwell.

Goodman, N., et al. 1963. "Variant Reactions to Physical Disabilities." *American Sociological Review*, 28: 429–35.

Goody, J. 1962. *Death, Property, and the Ancestors: A Study of Mortuary Custom Among the LoDagaa*. Stanford, Calif.: Stanford University Press.

Gordon, Aharon David. 1969. "Some Observations," in *The Zionist Idea*, ed. Arthur Herzberg, pp. 20–35. New York: Basic Books.

Gordon, N., and R. Marton, eds. 1995. *Torture: Human Rights, Medical Ethics and the Case of Israel*. London: Zed Books.

Gould, J. L. 1982. *Ethology*. New York: Norton.

Grinberg, Lev-Luis. 1994. "A Theoretical Framework for the Analysis of the Israeli Palestinian Peace Process," *Revue Internationale de Sociologie*, n.s. 1, 68–89.

Griswold, C. 1986. "The Vietnam Veterans Memorial and the Washington Mall: Philosophical Thoughts on Political Iconography," *Critical Inquiry*, 12.

Grover, K. 1989. *Fitness in American Culture: Images of Health, Sport, and the Body*. Amherst: University of Massachusetts Press.

Gvati, C. 1985. *A Hundred Years of Settlement: The Story of Jewish Settlement in The Land of Israel*. Jerusalem: Keter.

Haffter, C. 1968. "The Changeling: History and Psychodynamics of Attitudes to Handicapped Children in European Folklore," *Journal of the History of Behavioural Sciences*, 4: 55–63.

Halbwachs, Morris. 1980. *The Collective Memory*. New York: Harper & Row.

Hallam, E., J. Jockey, and G. Howarth. 1999. *Beyond the Body: Death and Social Identity*. New York: Routledge.

Handelman, Don, and Elihu Katz. 1990. "State Ceremonies of Israel—Remembrance Day and Independence Day," in *Models and Mirrors*, ed. Don Handelman, pp. 191–233. Cambridge: Cambridge University Press.

Haraway, Donna. 1989. "The Biopolitics of Postmodern Bodies: Determinations of Self in Immune System Discourse," *Differences*, 1, 1: 3–43.

Harvey, David. 1989. *The Condition of Postmodernity*. Oxford: Blackwell.

Hazan, Haim. 1994. *Old Age: Constructions and Deconstructions*. Cambridge: Cambridge University Press.

Heilman, S. 1983. *The People of the Book*. Chicago: University of Chicago Press.

Herman, Edward S. 1992. *Beyond Hypocrisy: Decoding the News in an Age of Propaganda*. Boston: South End Books.

Herman, Edward S., and Noam Chomsky. 1988. *Manufacturing Consent: The Political Economy of the Mass Media*. New York: Pantheon Books.

Herman, Simon N. 1970. *Israelis and Jews: The Continuity of an Identity*. New York: Random House.

Hever, Channan. 1995. "The Poetry of the Nation's Body: Women Poets in the War of Independence," *Téoria U'bikoret*, 7: 199–225 (in Hebrew)

Higonnet, M., et al. 1987. *Behind the Lines: Gender and the Two World Wars*. New Haven, Conn.: Yale University Press.

Hiss, Jehuda, and Tzipy Kahana. 1994. "The Fatalities of the *Intifada* (Palestinian Uprising): The First Five Years," *Journal of Forensic Science Society*, 34: 225–30.

———. 1996. "Medicolegal Investigation of Death in Custody: A Postmortem Procedure for Detection of Blunt Force Injuries," *American Journal of Forensic Medicine & Pathology*, 17, 4: 312–14.

Hochschild, Arlie, with Anne Maachung. 1989. *The Second Shift*. New York: Avon Books.

Holm, J. 1992. *Women in the Military: An Unfinished Revolution*. Novato, Calif.: Presidio.

Horowitz, Dan, and Baruch Kimmerling. 1974. "Some Social Implications of Military Service and the Reserve System in Israel," *Archives Européennes de Sociologie*, 15: 262–76.

Horowitz, Dan, and Moshe Lissak. 1978. *Origins of the Israeli Polity: Palestine Under the Mandate*. Chicago: University of Chicago Press.

Howes, R. H., and M. R. Stevenson, eds. 1993. *Women and the Use of Military Force*. London: Lynne Rienner.

Hunt, D. 1972. *Parents and Children in History*. New York: Harper & Row.

Israelashvili, Moshe. 1992. "Counseling in the Israeli High School: Particular Focus on Preparation for Military Recruitment," *International Journal for the Advancement of Counseling*, 15: 175–86.

Ivry, Tzipi. 1999. "Reproduction as Martial Art." Paper presented at the annual conference of the International Institute of Sociology, Tel Aviv, June. [Part of "Theories of Pregnancy in Japan and Israel," Ph.D. diss., Dept. of Sociology and Anthropology, Hebrew University of Jerusalem.]

Izraeli, Dafna. 1997. "Gendering Military Service in the Israeli Defense Forces," *Israel Social Science Research*, 12, 1: 129–66.

Jadresic, A. 1980. "Doctors and Tortures," *Journal of Medical Ethics*, 6: 124–27.

Jaggar, A., and S. Bordo. 1989. *Gender/Body/Knowledge*. New Brunswick, N.J.: Rutgers University Press.

Kahn, Susan M. 1997. "Reproducing Jews: The Social Uses and Cultural Meanings of the New Reproductive Technologies in Israel." Ph.D. diss., Dept. of Anthropology, Harvard University.

———. 2000. *Reproducing Jews: A Cultural Account of Assisted Conception*. Durham, N.C.: Duke University Press.

Kalisch, P. A., and M. Scobey. 1983. "Female Nurses in American Wars: Helplessness Suspended for the Duration," *Armed Forces and Society*, 9, 2: 215–45.

Kaplan, Steven. 1992. *The Beta Israel (Falasha) in Ethiopia*. New York: New York University Press.

Karton-Blum, Rut. 1996. "Fear of Isaac: Sacrifice Poems as a Case Study of the New Hebrew Poetry," in *Myth and Memory*, ed. D. Ohana and R. Wistrich, pp. 231–48. Tel Aviv: Hakibbutz Hame'uchad (in Hebrew).

Katriel, Tamar. 1986. *Talking Straight: Dugri Speech in Israeli Sabra Culture*. Cambridge: Cambridge University Press.

———. 1991. "Gibush: The Crystallization Metaphor in Israeli Cultural Semantics," in Katriel, *Communal Webs: Culture and Communication in Israel*, chap. 2. Albany: State University of New York Press.

Katz, Elihu, and Shmuel N. Eisenstadt. 1973. "Some Sociological Observations on the Response of Israeli Organizations to New Immigrants," in *Bureaucracy and the Public*, ed. E. Katz and B. Danet. New York: Basic Books.

Katz, Elihu, and Michael Gurevitch. 1976. *The Secularization of Leisure: Culture and Communication in Israel*. London: Faber & Faber.

Kellner, D. K. 1992. *The Persian Gulf TV War*. Boulder, Colo.: Westview Press.

Kimmerling, Baruch. 1974. "Anomie and Integration in Israeli Society and the Salience of the Israeli-Arab Conflict," *Studies in Comparative International Development*, 9, 3: 64–89.

———. 1985. *The Interrupted System*. New Brunswick, N.J.: Transaction.

———. 1992. "Sociology, Ideology, and Nation-Building: The Palestinians and Their Meaning in Israeli Sociology," *American Sociological Review*, 57: 446–60.

————. 1997. "A Moment of Solidarity," *Ha'arretz*, Feb. 13, p. B2.

Klepsch, M., and S. L. Logie. 1982. Children Draw and Tell: An Introduction to the Uses of Children's Human Figure Drawings. New York: Bruner & Mazel.

Kotkov, B., and M. Goodman. 1953. "The Draw-a-Person of Obese Women," *Journal of Clinical Psychology*, 9: 362–64.

Laqueur, Walter. 1972. *A History of Zionism.* New York: Basic Books.

————. 1990. *Making Sex, Body and Gender from the Greeks to Freud.* Cambridge, Mass.: Harvard University Press.

Lash, C., and J. Friedman, eds. 1992. *Modernity and Identity.* Oxford: Blackwell.

Levinger, E. 1993. *War Memorials in Israel.* Tel Aviv: Hakibbutz Hame'uchad (in Hebrew).

Levy-Schreiber, Edna, and Eyal Ben-Ari. 1998. "Body-Building, Character-Building and Nation-Building: Gender and Military Service in Israel," in *Military and Militarism in Israeli Society,* ed. Levy-Schreiber and Ben-Ari. Albany: State University of New York Press.

Lewis, B. 1975. *History Remembered, Recovered, Invented.* Princeton, N.J.: Princeton University Press.

Liebes, Tamar. 1997. *Reporting the Arab-Israeli Conflict: How Hegemony Works.* London: Routledge.

————. 1998. "Disaster Marathons: A Danger to the Democratic Process?," in *Media, Culture and Identity,* ed. T. Liebes and J. Curran. London: Routledge.

————. n.d. "Telling Moments: Public Discourse and 'Disaster Marathons.' " Unpublished manuscript, in Smart Institute of Communications, Hebrew University of Jerusalem.

Liebes, Tamar, and Yoram Peri. 1997. "Electronic Journalism in Segmented Societies," *Political Communication,* 15: 27–44.

Lieblich, Amiya. 1979. *Tin Soldiers on Jerusalem Beach.* New York: Schocken Books.

Lieblich, Amiya, and M. Perlow. 1988. "Transition to Adulthood During Military Service," *Jerusalem Quarterly,* 46: 40–76.

Liebman, Charles, and Eliezer Don-Yehia. 1983. *Civil Religion in Israel: Traditional Judaism and Political Culture in the Jewish State.* Berkeley: University of California Press.

Lienhardt, G. *Social Anthropology.* 2d ed. London: Oxford University Press.

Lifton, Robert J. 1977. "The Sense of Immortality: On Death and the Continuity of Life," in *Death in America,* ed. D. Stannard. Philadelphia: University of Pennsylvania Press.

————. 1979. The Broken Connection: On Death and the Continuity of Life. New York: Simon & Schuster.

Linke, Uli. 1999. *German Bodies: Race and Representation After Hitler.* New York: Routledge.

Lissak, Moshe, and Dan Horowitz. 1989. *Trouble in Utopia.* Albany: State University of New York Press.

Livne, Moshe. 1977. " 'Our Israel'—The Rise and Fall of a Protest Movement," M.A. thesis, Dept. of Sociology and Anthropology, Tel Aviv University (in Hebrew).

Lock, M. 1993. "Cultivating the Body: Anthropology and the Epistemologies of Bodily Practices and Knowledge," *Annual Review of Anthropology*, 22: 133–55.

Lomsky-Feder, Edna. 1994. "Patterns of Participation in War and the Constructions of War in the Life Course: Life Stories of Israeli Veterans from the Yom Kippur War." Ph.D. diss., Dept. of Sociology, Hebrew University of Jerusalem (in Hebrew).

———. 1998. *As If There Was No War*. Jerusalem: Magnes Press (in Hebrew).

Lomsky-Feder, Edna, and Eyal Ben-Ari. 1999. "From 'The People in Uniform' to 'Different Uniforms for the People': Professionalism, Diversity and the Israeli Defense Forces," in *Managing Diversity in the Armed Forces*, ed. J. Soeters and J. van der Meulen, pp. 157–86. Tilbourg: Tilbourg University Press.

Lutz, Catherine. 1988. *Unnatural Emotions*. Chicago: University of Chicago Press.

MacArthur, J. R. 1992. *Second Front: Censorship and Propaganda in the Gulf War*. New York: Hill &Wang.

MacGregor, F. C. 1980. *Transformations and Identity: The Face and Plastic Surgery*. Oak Grove, Ill.: Eterna Press.

Maoz, Rivka. 1988. "The Transformations of the Blorit in Israeli Literature," Moznaim, 5–6: 50–57 (in Hebrew).

Martin, Emily. 1987. *The Woman in the Body: A Cultural Analysis of Reproduction*. Boston: Beacon Press.

———. 1990. "The End of the Body?," *American Ethnologist*, 18: 121–38.

———. 1991. "The Egg and the Sperm: How Science Has Constructed a Romance Based on Stereotypical Male-Female Roles," *Signs*, 16, 3: 1–18.

———. 1994. *Flexible Bodies: The Role of Immunity in American Culture from the Days of Polio to the Age of AIDS*. Boston: Beacon Press.

Marx, Tzvi. 1993. "Halakha and Handicap: Jewish Law and Ethics on Disability." Unpublished manuscript, Jerusalem. [All rights reserved to the author.]

de Mause, L., ed. 1976. *The History of Childhood*. London: Souvenir Press.

McIntyre, C. 1990. *Monuments of War: How to Read a War Memorial*. London: Robert Hale.

Mennel, S. 1991. "On the Civilizing of Appetite," in *The Body: Social Process and Cultural Theory*, ed. M. Featherstone et al., pp. 126–57. London: Sage.

Merleau-Ponty, Maurice. 1962. *The Phenomenology of Perception*, tr. Colin Smith. London: Routledge & Kegan Paul.

Metzner-Licht, B., et al. 1980. "Pregnancy Termination Committees—The Reflection of Social Problems from the Point of View of Social Workers," *Society and Welfare*, 3c (in Hebrew).

Middleton, D., and D. Edwards, eds. 1990. *Collective Remembering*. London: Sage.

Miron, Dan. 1992. *In Front of the Silent Brother: Studies in the Poetry of the War of Independence*. Jerusalem: The Open University and Keter (in Hebrew).

Moore, Sally Anne, and Barbara Myerhoff, eds. 1977. *Secular Ritual.* Essen: Van Gorcum.

Mosse, George. 1979. "National Cemeteries and National Revival: The Cult of the Fallen Soldiers in Germany," *Journal of Contemporary History,* 14.

———. 1990. *Fallen Soldiers: Reshaping the Memory of the World Wars.* Oxford: Oxford University Press.

———. 1996. *The Image of Man: The Creation of Modern Masculinity.* New York: Oxford University Press.

Mowlana, H., G. Gerbner, and H. Schiller, eds. 1992. *Triumph of the Image: The Media's War in the Persian Gulf—A Global Perspective.* Boulder, Colo.: Westview Press.

NAS (National Academy of Sciences Committee on Human Rights and Institute of Medicine Committee on Health and Human Rights). 1992. *Scientists and Human Rights in Guatemala: Report of a Delegation.* Washington, D.C.: Govt. Printing Office.

Nave, Hana. 1993. *Captives of Mourning.* Tel Aviv: Hakibbutz Hame'uchad (in Hebrew).

Navon, Yitzhak, et. al. 1996. *Report of the Commission Investigating the Ethiopian Immigrants' Blood Donations Affair.* Jerusalem: Government of the State of Israel (in Hebrew).

Nightingale, Florence. 1882. "Training of Nurses and Nursing the Sick," in *Dictionary of Medicine,* ed. D. Quain. London: Spottiswoode.

Norman, Elizabeth. 1990. *Women at War: The Story of 50 Military Nurses Who Served in Vietnam.* Philadelphia: University of Pennsylvania Press.

Nossek, H. 1994. "The Narrative Role of the Holocaust and the State of Israel in the Coverage of Terrorist Events in the Israeli Press," *Journal of Narrative and Life History,* 4, 1–2: 119–35.

Offrat, Gideon. 1990. *Earth, Man, Blood: The Myth of the Pioneer and the Ritual of Earth in the Settlement Drama of Eretz-Israel.* Tel Aviv: Cherikover (in Hebrew).

O'Neill, John. 1989. *The Communicative Body.* Evanston, Ill.: Northwestern University Press.

Ortner, S. 1973. "On Key Symbols," *American Anthropologist,* 75: 1338–1346.

Paine, Robert. 1993. "Israel: The Making of Self in the 'Pioneering' of the Nation," *Ethnos,* 58, 3-4: 222–40.

Paletz, David L., and Alex P. Schmid, eds. 1992. *Terrorism and the Media.* Newbury Park, Calif.: Sage.

Palgi, Philis. 1974. *Death, Mourning, and Bereavement in Israeli Society During War.* Jerusalem: Ministry of Health (in Hebrew).

Palgi, Philis, and Henry Abramovitz. 1984. "Death: A Cross-Cultural Perspective," *Annual Review of Anthropology,* 13: 385–417.

Palgi, Philis, and Jacob Durban. 1995. "The Role and Function of Collective Representations for the Individual During the Mourning Process: The Case of a War-Orphaned Boy in Israel," *Ethos*, 23, 2: 223–43.

Pandolfo, S. 1989. "Detours of Life: Space and Bodies in a Moroccan Village," *American Ethnologist*, 16, 1: 3–24.

Pollock, L. 1983. *Forgotten Children: Parent-Child Relations from 1500 to 1900*. Cambridge: Cambridge University Press.

Porat, Dina. 1986. *Leadership in Conflict*. Tel Aviv: Am-Oved (in Hebrew).

Rabinow, Paul. 1984. *The Foucault Reader*. New York: Pantheon Books.

Racault, J. M. 1986. "Corps utopiques, utopies du corps," in *Collectif: Pratiques du Corps, Medécine, Hygiène, Alimentation, Sexualité*. Paris: Didier.

Ram, Uri. 1989. "Civic Discourse in Israeli Sociological Thought," *International Journal of Politics, Culture and Society*, 3, 2: 255–72.

———. 1993a. "Emerging Modalities of Feminist Sociology in Israel," *Israel Social Science Research*, 8, 2: 51–76.

———. 1993b. "The Colonization Perspective in Israeli Sociology: Internal and External Comparisons," *Journal of Historical Sociology*, 6, 3: 327–50.

Reich, A. 1972. "Changes and Developments in the Passover Haggadot of the Kibbutz Movement." Ph.D. diss., University of Texas, Austin.

Richardson, Laurel. 1997. *Fields of Play: Constructing an Academic Life*. New Brunswick, N.J.: Rutgers University Press.

Robin, R., B. Pacifici, and B. Schwartz. 1991. "The Vietnam Veterans' Memorial: Commemorating a Difficult Past," *American Journal of Sociology*, 97, 2: 376–420.

Roniger, Louis, and Michael Feige. 1992. "From Pioneer to Freier: The Changing Models of Generalized Exchange in Israel," *Archives of European Sociology*, 33: 280–307

Rorty, Richard. 1986. *The Mirror of Nature*. Cambridge, Mass.: Harvard University Press.

Rosaldo, Michelle. 1984. "Towards an Anthropology of Self and Feeling," in *Culture Theory*, ed. R. Schweder and R. LeVine, pp. 137–54. Cambridge: Cambridge University Press.

Rosen, F. 1972. "Autopsy in Jewish Law and the Israeli Autopsy Controversy." *Studies in Torah Judaism (Modern Medicine and Jewish Law)*. Dept. of Special Publications, Yeshiva University.

Roumani, Maurice M. 1980. *From Immigrant to Citizen: The Contribution of the Army to National Integration in Israel*. The Hague: Foundation for the Study of Plural Societies.

Rubinstein, Amnon. 1977. "The Rise and Fall of the Mythological Sabra," in Rubenstein, *To Be a Free People*, chap. 6. Tel Aviv: Shoken (in Hebrew).

Ruddick, Sara. 1989. *Maternal Thinking*. New York: Ballantine Books.

Sagi, Michal. 1999. "Ethical Aspects of Genetic Screening in Israel," *Science in Context*, 11, 3–4: 419–29.

Salamon, Hagar. 1995. "Reflections of Ethiopian Cultural Patterns on the Beta Israel Absorption in Israel: The 'Barya' Case," in *Between Africa and Zion: Proceedings of the First International Congress of the Society for the Study of Ethiopian Jewry*, pp. 126–30. Jerusalem: Ben-Tzvi Institute.

Salyer, M., A. Jensen, and C. Borden. 1985. "Effects of Facial Deformities and Physical Attractiveness on Mother-Infant Bonding," in *Craniofacial Surgery: Proceedings of the First International Society of Cranio-Maxillo-Facial Surgery*.

Schaffer, H. R., ed. 1977. *Studies in Mother-Infant Interaction*. London: Academic Press.

Scheper-Hughes, Nancy. 1991. *Death Without Weeping*. Berkeley: University of California Press.

———. 2000. "The Global Traffic in Human Organs," *Current Anthropology*, 41, 2: 191–211.

Scheper-Hughes, Nancy, and Margaret Lock. 1987. "The Mindful Body: A Prolegomenon to Future Work in Medical Anthropology," *Medical Anthropology Quarterly*, 1, 1: 6–42.

Schiller, H. 1992. "Manipulating Hearts and Minds," in *Triumph of the Image: The Media's War in the Persian Gulf—A Global Perspective*, ed. H. Mowlana et al., pp. 22-29. Boulder, Colo.: Westview Press.

Schneider, David. 1980. *American Kinship: A Cultural Account*. 2d ed. Chicago: University of Chicago Press.

Schwartz, Barry. 1982. "The Social Context of Commemoration: A Study in Collective Memory," *Social Forces*, 61, 2: 374–462.

Schwartz, Barry, Yael Zerubavel, and B. M. Barnett. 1986. "The Recovery of Masada," *Sociological Quarterly*, 27, 2: 147–64.

Scrimshaw, S. 1984. "Infanticide in Human Populations," in *Infanticide*, ed. Hausfater and Hardy. New York: Aldine.

Seeman, Don F. 1997. "One People, One Blood: Religious Conversion, Public Health, and Immigration as Social Experience for Ethiopian-Israelis." Ph.D. diss., Dept. of Anthropology, Harvard University.

Segal, M. W. 1995. "Women's Military Roles Cross-Nationally: Past, Present, and Future," *Gender and Society*, 9, 6: 757–75.

Segev, Tom. 1989. "Catalog of Macabre Israeli Tombstones," *Ha'arretz* (in Hebrew).

———. 1991. *The Seventh Million*. New York: Maxwell-Macmillan.

Sered, Susan. 1990. "Women, Religion and Modernization: Tradition and Transformation Among Elderly Jews in Israel," *American Anthropologist*, 92: 306–18.

———. 1992. *Women and Religion*. Oxford: Oxford University Press.

Shafir, Gershon. 1996. "Israeli Decolonization and Critical Sociology," *Journal of Palestine Studies*, 25, 3: 23–35.

Shafir, Gershon, and Yoav Peled. 1996. "The Roots of Peacemaking: The Dynamics of

Citizenship in Israel, 1948–93," *International Journal of Middle East Studies*, 28, 3: 391–413.

———. 1998. "Citizenship and Stratification in an Ethnic Democracy," *Ethnic and Racial Studies*, 21, 3: 408–27.

———. 1996. "The Roots of Peacemaking: The Dynamics of Citizenship in Israel, 1948–93," *International Journal of Middle East Studies*, 28, 3: 391–413.

Shaham, N., and T. Ra'anan, eds. 1991. *War in the Gulf: A Collection of Essays*. Tel Aviv: Sifriat Poalim (in Hebrew).

Shamgar-Handelman, Leah. 1986. *Israeli War Widows: Beyond the Glory of Heroism*. New York: Bergin & Garvey.

Shapira, Rina, and F. Hayman. 1991. "Solving Educational Dilemmas by Parental Choice: The Case of Israel," *International Journal of Educational Research*, 15: 3–14.

Shapira, Rina, and Hana Herzog. 1984. "Understanding Youth Culture Through Autograph Books," *Journal of American Folklore*, 97 (386): 443–60.

Shavit, Ya'akov. 1984. *From Hebrew to Canaanite*. Tel Aviv: Domino (in Hebrew).

Shenkar, Y. 1996. "Is It Justified to Kill a Fetus for Medical Reasons?," *Medicine*, 131, c–d: 101–3 (in Hebrew).

Shilling, Chris. 1993. *The Body and Social Theory*. London: Sage.

Shokeid, Moshe. 1988. *Children of Circumstances*. Ithaca, N.Y.: Cornell University Press.

Shokeid, Moshe, and Shlomo Deshen. 1974. *The Predicament of Homecoming: Cultural and Social Life of North African Immigrants in Israel*. Ithaca, N.Y.: Cornell University Press.

Shuval, Judith. 1992. *Social Dimensions of Health: The Israeli Experience*. New York: Praeger.

Silverstein, A., and H. Robinson. 1956. "The Representation of Orthopedic Disability in Children's Figure Drawing," *Journal of Consulting Psychology*, 20: 333–41.

Sion, Liora. 1997. "Images of Manhood Among Combat Soldiers: Military Service in the Israeli Infantry as a Rite of Initiation from Youth to Adulthood." *Shayne Working Paper* no. 3, Faculty of Social Sciences, Hebrew University of Jerusalem (in Hebrew).

Sivan, Imanuel. 1986. *The 1948 Generation: Myth, Profile and Memory*. Tel Aviv: Ministry of Defense Publications (in Hebrew).

Smith, C. 1991. "From 1960s' Automation to Flexible Specialization: A Deja-vu of Technological Panaceas," in *Farewell to Flexibility?*, ed. A. Pollert. London: Blackwell.

Snitow, Ann. 1992. "Feminism and Motherhood: An American Reading," *Feminist Reviews*, 40: 33–52.

Spinetta, J., H. McLaren, R. Fox, and S. Sparta. 1981. "The Kinetic Family Drawing in Childhood Cancer," in *Living with Childhood Cancer*, ed. J. Spinetta and P. Spinetta, pp. 86–120. St. Louis, Mo.: C. V. Mosby.

Spiro, Melford. 1965. *Children of the Kibbutz*. New York: Shocken Books.

Stern, M. 1992. "Reflections on the Gulf War," *Sichot*, 7, 1 (in Hebrew).

Stone, G. P. 1962. "Appearance and the Self," in *Human Behavior and Social Process*, ed. A. M. Rose, pp. 86–118. Boston: Houghton-Mifflin.

Synnott, A. 1992. "Tomb, Temple, Machine and Self: The Social Construction of the Body," *British Journal of Sociology*, 43, 1: 79–110.

Tait, C., and R. Ascher. 1955. "Inside of the Body Test," *Psychosomatic Medicine*, 17: 139–48.

Taylor, L., ed. 1989. "The Uses of Death in Europe," *Anthropological Quarterly*, 62, 4.

Taylor, P. M. 1992. *War and the Media: Propaganda and Persuasion in the Gulf War*. Manchester, Eng.: Manchester University Press.

Tedeschi, L. 1984. "Methodology in Forensic Medicine: Documentation of Human Rights Abuses," *American Journal of Forensic Medicine and Pathology*, 5, 4: 301–16.

Theweleit, Klaus. 1989. *Male Fantasies*. Vol. 2: *Male Bodies: Psychoanalyzing the White Terror*, tr. Erica Carter and Chris Turner. Minneapolis: University of Minnesota Press.

Timen, Elli. 1999. "Being One Body: Surrogate Motherhood in Israel." Paper presented at the annual conference of the International Institute of Sociology, Tel Aviv, June.

Tress, Madeleine. 1994. "Does Gender Matter?," *Books on Israel*, 3: 89–105.

Turner, Bryan S. 1991a. "The Discourse of Diet," in *The Body: Social Process and Cultural Theory*, ed. M. Featherstone et al., pp. 157–70. London: Sage.

———. 1991b. "Missing Bodies—Towards a Sociology of Embodiment, *Sociology of Health and Illness*, 13, 2: 265–73.

———. 1992. *Regulating Bodies: Essays in Medical Sociology*. London: Routledge.

Turner, Victor. 1977. *The Ritual Process*. Ithaca, N.Y.: Cornell University Press.

Van Gennep, Arnold. 1960 [1908]. *The Rites of Passage*. Chicago: University of Chicago Press.

Vital, David. 1975. *The Origins of Zionism*. Oxford: Oxford University Press.

Wagaw, Teshome. 1993. *For Our Souls: Ethiopian Jews in Israel*. Detroit: Wayne State University Press.

Warner, W. Lloyd. 1959. *The Living and the Dead: A Study of the Symbolic Life of the Americans*. New Haven, Conn.: Yale University Press.

Weimann, Gabriel. 1990. " 'Redefinition of Image': The Impact of Mass-Mediated Terrorism," *International Journal of Public Opinion Research*, 2, 1: 16–29.

Weingrod, Alex. 1995. "Dry Bones: Nationalism and Symbolism in Contemporary Israel," *Anthropology Today*, December.

Weiss, Meira. 1978. "The Bereaved Parent's Position: Aspects of Life Review, Rebellion and Self-Fulfillment." M.A. thesis, Dept. of Anthropology, Tel Aviv University (in Hebrew).

———. 1989. "The Bereaved Parent's Position: Aspects of Life Review and Self-Fulfillment," *Current Perspectives on Aging and the Life Cycle*, 3: 269–80.

————. 1994. *Conditional Love*. New York: Bergin & Garvey.

————. 1997a. "Signifying the Pandemics: Metaphors of AIDS, Cancer and Heart Disease," *Medical Anthropology Quarterly*, 11, 4: 1–21.

————. 1997b. "War Bodies–Hedonist Bodies: Dialectics of the Collective and the Individual in Israeli Society," *American Ethnologist*, 24, 4: 1–20.

————. 1997c. "For Doctors' Eyes Only: Medical Records in Two Israeli Hospitals," *Culture, Medicine, and Psychiatry*, 21, 3: 283–302.

————. 1997d. "Bereavement, Commemoration, and Collective Identity in Contemporary Israeli Society," *Anthropological Quarterly*, 70, 2: 91–101.

————. 1998a. "Narratives of Embodiment: The Discursive Formulation of Multiple Bodies," *Semiotica*, 118, 3–4: 239–60.

————. 1998b. "Parents' Rejection of Their Appearance-Impaired Newborns: Some Critical Observations Regarding the Social Myth of Bonding," *Marriage and Family Review*, 27, 3–4: 191–209.

————. 1998c. "Conditions of Mothering: The Bio-Politics of Falling in Love with Your Child," *Social Science Journal*, 35, 1: 87–105.

Werman, Robert. 1993. *Notes from a Sealed Room: An Israeli View of the Gulf War*. Carbondale: Southern Illinois University Press.

Wertz, Dorothy. 2000. "Eugenics Is Alive and Well: A Survey of Genetic Professionals Around the World," *Science in Context*, 3–4: 493–510.

Whaley, J., ed. 1988. *Mirrors of Mortality: Studies in the Social History of Death*.

Wittebols, James. 1992. "Media and the Institutional Perspective: U.S. and Canadian Coverage of Terrorism," *Political Communication*, 9, 4: 267–78.

Witztum, E., and R. Malkinson. 1993. "Bereavement and Commemoration: The Dual Faces of National Myth," in *Loss and Bereavement in the Jewish Society of Israel*, ed. Malkinson et al., pp. 231–95 (in Hebrew).

Wizeltir, Meir. 1980. "A Longitudinal Analysis of the Poetry of Natan Zach," *Siman Kria'a*, 10 (in Hebrew).

Wysocki, A., and E. Wysocki. 1977. "Human Figure Drawings of Sex Offenders," *Journal of Clinical Psychology*, 33, 1: 278–84.

Yafe, A. B. 1989. *Written in Tashach* (1948), in *Anthology of Poetry and Prose Written During Israel's War of Independence*, ed. A. B. Yafe. Tel Aviv: Reshafim (in Hebrew).

Yanai, Nathan. 1996. "The Citizen as Pioneer: Ben-Gurion's Concept of Citizenship," *Israel Studies*, 1, 1: 127–43.

Young, James E. 1990. "When a Day Remembers: A Performative History of Yom Hashoah," *History and Society*, 2, 2: 54–75.

————. 1991. *Writing and Rewriting the Holocaust: Narrative and the Consequences of Interpretation*. Bloomington: Indiana University Press.

Yuchtman-Yaar, Ephraim, Yochanan Peres, and Dina Goldberg. 1994. "Doing Research Under Missiles: Israeli Morale During the Gulf War." Unpublished paper, in Dept. of Sociology, Tel Aviv University.

Yuval-Davis, Nira. 1985. "Front and Rear: The Sexual Division of Labor in the Israeli Army," *Feminist Studies*, 11, 3: 648–75.

Zerubavel, Yael. 1980. "The Last Stand: The Transformations of Symbols in Modern Israel." Ph.D. diss., University of Pennsylvania.

———. 1990. "The Historic, the Legendary, and the Incredible: Invented Tradition and Collective Memory in Israel," in *Commemorations: The Politics of National Identity*, ed. J. Gillis, pp. 105–27. Princeton, N.J.: Princeton University Press.

———. 1991. "New Beginnings, Old Past: The Collective Memory of Pioneering in Israeli Culture," in *New Perspectives on Israeli History: The Early Years of the State*, ed. L. Silberstein, pp. 193-215. New York: New York University Press.

———. 1996. *Recovered Roots*. Albany: State University of New York Press.

Zola, Irvin Kenneth, ed. 1992. *Disability Studies Quarterly*, special issue: "The Body" (vol. 12, no. 2).

INDEX

AIDS, and body image, 57
Ahad Ha'am, 20, 148fn14
Amir, Yigal, 132–34
Aronoff, M., 86
Appearance, children with impaired, 32–38
Ashkenazi and Mizrachi, 29

Beuf, A., 33
Ben-Ari, Eyal, 44
Biale, David, 3–4, 21
Body: theories of, in sociology and anthro-
pology, 1–5, 9–14, 146fn2; regulation of,
9–10; body image, 12–14, 41, 52–53; the
"body politic," 11–12, 16–18, 131; and femi-
nist perspectives, 11–12, 147fn10,11; and
gender, 14–16; of fallen soldiers, 60–61
BodyTalk, 122–23, 131
Bourdieu, Pierre, 12
Boyarin, Daniel, 3–4

Collectivism, 84–86; and Zionism, 5–6,
22–23; and body imagery, 6, 9, 22; and
Fordism, 22; in the Gulf War, 55–56
Csordas, Thomas, 146fn3

Diaspora, as degeneration, 24–25, 94
Disabilities, people with, 88–90
Douglas, Mary, 10, 73, 93
Durkheim, Emile, 10

Eilberg-Schwartz, H., 3–4
Embodiment, 10–11, 22–24

Ethiopian-Israelis, 62
Eugenics, 2–3, 28–32

Family, as agent of socialization, 45–46
Fetal defects, 31–32
Foucault, Michel, 10–11, 153fn12
Fordism (as collectivism in the United
States), 22
Frank, Arthur, 12
Frank, Gelya, 154fn14
Friedmann, G., 4

Gender, 116; and Zionism, 15–16; and status
conflict, 113, 116; and soldiering, 99–101,
111–15
Genetic testing and screening, 28–32
German body, 147fn8
Gerzon, M., 27
Gluzman, Michael, 4, 95
Gordon, A. D., 1, 21

Herzl, Theodor, 4, 15, 20
Hever, Chanan, 96–99
Holocaust, 25, 148fn16

Jabotinsky, Ze'ev, 1–2

Kahn, Susan, 40–41
Kimmerling, Baruch, 65–66

Levin, Hanoch, 73–74, 153fn9
Liebes, Tamar, 119–20, 127–28

Lock, Margaret, 9–10
Lomsky-Feder, Edna, 47

Manhood, 44, 46
Martin, Emily, 6, 12–14
Masada myth, 85, 112
Masculinity, 14–16, 95, 150fn11
Maternal thinking, 115
Medicine: and pregnancy, 28–31; and new-
    borns, 34–35; forensic, 58–60, 129; and
    military, 60–61; autopsies, 151fn15
Memorial Day, 67–69, 84
Militarism, 6; pre-socialization in high
    school, 42–43; and hazing, 48–51; aggres-
    sion training, 49; and homoeroticism, 51;
    fallen soldiers, 67–74; and sacrifice,
    68–70
Military occupation, 16
Mosse, George, 49, 59, 75
Motherhood, 40–41

Nordau, Max, 1, 32, 94

Organ harvesting, 151fn18

Pioneer (*halutz*), 5–6, 18–21, 56
Poetry: and the chosen body, 70–72; in the
    War of Independence, 96–98

Rabin, Yitzhak, 131–34
Ram, Uri, 17
Reproductive technologies, 40–41
Roth, Philip, 3

Sabra, 18–26, 70–71
Scheper-Hughes, Nancy, 39
Segev, Tom, 75–77, 83
Shafir, Gershon, 17–18
Shapira, Anita, 4
Shilling, C., 11
Soldiering, 111–12. *See also* militarization
Stigmatization, 150fn9,10

Tabenkin, Yitzhak, 24
Trumpeldor, Yosef, 22, 85

Van Gennep, Arnold, 10

War memorials, 75, 81, 121
Warner, F. Lloyd, 84
Women's Corps (CHEN), 96

Yemenite children affair, 61

Zionism: and the body, 1–2; and Eros, 3–4,
    21; and gender inequality, 15–16; and colo-
    nialism, 16–18

CONTRAVERSIONS

JEWS AND OTHER DIFFERENCES

Stephen D. Moore, *God's Beauty Parlor: And Other Queer Spaces in and around the Bible*

Rela Mazali, *Maps of Women's Goings and Stayings*

Shelly Matthews, *First Converts: Rich Pagan Women and the Rhetoric of Mission in Early Judaism and Christianity*

Menachem Lorberbaum, *Politics and the Limits of Law: Secularizing the Political in Medieval Jewish Thought*

Gabriella Safran, *Rewriting the Jew: Assimilation Narratives in the Russian Empire*

Galit Hasan-Rokem, *Web of Life: Folklore in Rabbinic Literature*

Charlotte Elisheva Fonrobert, *Menstrual Purity: Rabbinic and Christian Reconstructions of Biblical Gender*

James A. Matisoff, *Blessings, Curses, Hopes, and Fears: Psycho-Ostensive Expressions in Yiddish*, second edition

Benjamin Harshav, *The Meaning of Yiddish*

Benjamin Harshav, *Language in Time of Revolution*

Amir Sumaka'i Fink and Jacob Press, *Independence Park: The Lives of Gay Men in Israel*

Alon Goshen-Gottstein, *The Sinner and the Amnesiac: The Rabbinic Invention of Elisha ben Abuya and Eleazar ben Arach*

Bryan Cheyette and Laura Marcus, eds., *Modernity, Culture, and 'the Jew'*

Benjamin D. Sommer, *A Prophet Reads Scripture: Allusion in Isaiah 40–66*

Marilyn Reizbaum, *James Joyce's Judaic Other*

Printed and bound by CPI Group (UK) Ltd, Croydon, CR0 4YY

16/04/2025

14658403-0001